STRANGE WEATHER

THE HAYMARKET SERIES

Editors: Mike Davis and Michael Sprinker

The Haymarket Series offers original studies in politics, history and culture, with a focus on North America. Representing views across the American left on a wide range of subjects, the series will be of interest to socialists both in the USA and throughout the world. A century after the first May Day, the American left remains in the shadow of those martyrs whom the Haymarket Series honours and commemorates. These studies testify to the living legacy of political activism and commitment for which they gave their lives.

STRANGE WEATHER

Culture, Science, and Technology in the Age of Limits

ANDREW ROSS

VERSO
London · New York

First published by Verso 1991

Reprinted 1996

Verso
UK: 6 Meard Street, London W1V 3HR
USA: 180 Varick Street, New York, NY 10014-4606

Verso is the imprint of New Left Books

British Library Cataloguing in Publication Data

Ross, Andrew
Strange weather : culture, science and technology
in the age of limits.
I. Title
306

ISBN 0-86091-354-6
ISBN 0-86091-567-0 (paperback)

Library of Congress Cataloging-in-Publication Data

Ross, Andrew, 1956–
Strange weather : culture, science, and technology in the age of
limits / Andrew Ross.
p. cm.
Includes bibliographical references and index.
ISBN 0-86091-354-6 (hb). — ISBN 0-86091-567-0 (pb)
1. Science—Philosophy. 2. Science—Social aspects.
3. Technology—Philosophy. 4. Technology—Social aspects.
5. Social change. 6. Social problems. I. Title.
Q175.R5547 1991
501—dc20

Cover design by Tom Zummer. rilm stills
from *Peggy and Fred in Hell* by Leslie Thornton.
Frontispiece: 'The Baby Lenin Disk Drive',
Andrew Ross, mixed media, 1990

Typeset by York House Typographic Ltd, London
Printed in the U.S.A. by Courier Companies Inc.

CONTENTS

ACKNOWLEDGMENTS

This book is dedicated to all of the science teachers I never had. It could only have been written without them.

Thanks are also due to friends who have tolerated, and in some cases, encouraged my obsessions with the weather over the years.

Many people helped with material, information, and ideas. Joel Kovel, Meaghan Morris, Michael Sprinker, and Constance Penley all read complete drafts of the book in detail, and each offered invaluable responses and spirited comments. I am also grateful to members of the Bary Area editorial collective of *Socialist Review,* for an extended dialogue about the shape of one chapter, and to my companions on the *Social Text* collective for their collusion in the kind of cultural politics that *Strange Weather* advocates.

The bulk of this book was written while I was a fellow at Cornell University's Society for the Humanities. My thanks to everyone at the A.D. White House in Ithaca—Jonathan Culler, Mary Ahl, Aggie Sirrine, and Linda Allen—and also to Lucinda Rosenfeld, my research assistance at Cornell, and Jennifer Hayward, my research assistant at Princeton.

Earlier versions, or portions, of some of these chapters appeared in *Postmodern Culture, Critical Inquiry, Social Text,* and the *Socialist Review.* Thanks to University of Chicago Press for permission to reprint.

RS232
GENDER CHANGER
M/M

INTRODUCTION

While writing and researching much of this book I used the same photocopy machine from day to day. Above the machine, fixed to the wall, was the following notice:

> WARNING!
>
> *This machine is subject to breakdowns during periods of critical need*
>
> A special circuit in the machine called a "critical detector" senses the operator's emotional state in terms of how desperate he or she is to use the machine. The "critical detector" then creates a malfunction proportional to the desperation of the operator. Threatening the machine with violence only aggravates the situation. Likewise, attempts to use another machine may cause it also to malfunction. They belong to the same union. Keep cool and say nice things to the machine. Nothing else seems to work.
>
> *Never let anything mechanical know you are in a hurry*

These humorous words of warning were not drafted with this particular photocopy machine in mind. This is a commonplace notice to be found in many different workplaces, adjacent to many different machines—although I suspect it is more likely to be found in an office workplace than on a manufacturing shopfloor, or in the cockpit of a Stealth fighter–bomber. The warning also seems more applicable to a "smart" information machine than to a more inert mechanical technology. In personifying the machine as a unit of organized labor, sharing fraternal interests and union loyalties with other machines, the notice assumes a degree of evolved self-consciousness on the machine's part. Furthermore, it implies a relation of hostility, as if the machine's self-consciousness and loyalty to its own kind have inevitably led to resentment, conflict, and sabotage. In this respect, the notice reproduces typically anti-labor sentiments in the guise of daily office humor. To achieve its droll purpose, the warning exploits many traditional working-class fears and anxieties about technology that date from the time of the first mechanized workplaces. It is all the more ironic, then, that the machine is personified as a scheming, untrustworthy worker and that the reader is invited to share the management's viewpoint.

But the notice can also be read as a "warning" of a different kind. From this other perspective, it carries a message about the new smart machine, technologically equipped with the means of surveillance and control over the operators themselves. Programmed to detect and punish any evidence of hostility or maladjustment on the operator's part, the smart machine can now be seen as a management tool, enforcing the operator's reasonable behavior—"keep cool and say nice things"—and addressing him or her as a worker. The "union" to which the machine is loyal now represents the interests of owners and managers. Presented as friendly advice to a harried user— "nothing else seems to work"—this notice clearly carries a punitive warning about behavior that is out of line—"never let anything mechanical know you are in a hurry."

Humorous notices like this one are part of the daily culture of work environments that involve advanced technology. They are an element of the games that we play with technology, and the games that are played with us. The divided messages they generate are saturated with the psychology of regulated fear that has grown up around the use of such technologies, especially those developed partly to control worker behavior. It would be silly to imagine some conspiratorial editorial group employed precisely to exploit this psychology in penning such messages, and yet the consequences are often the same as if there were such a group. Whether the operator is addressed as management or as worker, the warning appeals finally to a kind of controlled reasonableness that accords well with the planned "rationalization" of the workplace through the development and installation of industrial, and now postindustrial, technologies— "nothing else seems to work."

No doubt we could learn even more about this warning through further analysis of its rhetoric and the messages it conveys. We only need so much analysis, however, to recognize that the range of fears and desires called up by modern technologies and played upon in this notice is much more complex, intimate, and mediated than any simple picture of conflict between "man and machine" of the sort that traditionally weighs the technology's potential against its dangers. There is no mistaking the graphic signs of class conflict, whether we see the machine as shifty worker or as management spy. But the degree of emotional conflict triggered by this friendly warning suggests the complexity with which we domesticate our relation to technology, always in connection with preexisting patterns of feelings and expectations.

For one thing, the warning is itself part of the technological environment; its rhetoric speaks to our "emotional state," while appealing to a framework of rules that finally may be no less prescriptive and every bit as rationalized as the catalog of instructions that accompanies each new machine. We are reminded, then, that technologies are much more than hardware objects or technical extensions of the human body. Technologies are also intentional linguistic processes; their ruling precepts are apparent even in the form of a crafty rhetoric that can divide our allegiances. We are accustomed to think of technology's precision methods as the result of a disciplined approach to rationalizing the object world, employing a scientific calculus of quantity and efficiency that is quite at odds with traditional ways—aesthetic, religious, ideological—of making sense of the world. But however remote, impersonal, or alienating these processes are, technologies are also fully lived and experienced in our daily actions and practices, and that is why it is important to understand technology not as a mechanical imposition on our lives but as a fully cultural process, soaked through with social meaning that only makes sense in the context of familiar kinds of behavior. Technologies cannot simply *determine* our behavior, although they are part of a persistent, and often coercive, dialogue about our manners.

Armchair speculation about the context-free "essence" of technology is common enough. All the more important, then, for me to acknowledge the contexts and limits of my own speculations. The photocopy machine is only one of the media technologies I regularly employ in my own work as a cultural critic and educator who does not consider himself especially technoliterate. The roll-call of names and models is the dull, quixotic sound of corporate poetry: a Toshiba T3100e portable personal computer; a souped-up IBM PC, with a Hayes 2000 modem; three printers—the Hewlett Packard DeskJet 500, the Brother HR-15XL, and an Epson FX-80; a Magnavox RK5565 television set; a Panasonic PV-2812 VCR; a Nikon N2020 AF camera; an Adler-Royal FX 915 fax machine, an IBM Wheelwriter 10 Series II typewriter; a Sony Soundrider CFS-W501 cassette recorder; a Sony Walkman WM-AF55; and an auto stereo system; in addition to film and video projection technologies that I use at my university workplace. Add to this the fact that about 62 cents from every tax dollar (a pre-Gulf War estimate by the War Resisters League) deducted from my salary helps to advance the science of making and maintaining destructive military technology or to pay for past, present, and future wars. The result is a graphic testimony to the substantial presence of technology in the cultural and economic life of a humanist intellectual, a species usually perceived as benignly remote from the technological world. Such an image, of course, is flatly anachronistic. Given the highly developed role of information technologies and bureaucracies in modern academic life, it has become almost rote in recent years to think of even the humanist intellectual as a knowledge worker whose "productivity" or "output" is strictly regulated by the systematic regime of "professional standards."

The maintenance of this pseudo-industrial system of cultural production has a social cost that goes beyond the "proletarianization" of knowledge workers. The warning above the photocopy machine was not the only such message in the office. Within a few feet of the photocopy machine stood a bin for recyclable paper (an increasingly standard feature of the office landscape), communicating an altogether different message about the technology of knowledge production. The authoritative presence of such bins reminds us of the troubled physical underpinnings of a technological system designed to ensure the rapid circulation of knowledge. The loud, colorful graphics on the side of the bin allude to the finitude of natural resources with a timely sense of urgency that contradicts the otherwise "magical" promise of infinite reproducibility offered by the photocopy machine. The message is a corrective. While it is important to consider the political ecology of such machines' architecture—their capacity to democratize information, or to regulate access to information—the recycling sign asks us to remember that they are also elements in a larger physical ecology that does not reproduce or renew its resources according to the same technological principles as the machines.

In many ways, the dialogue between these two messages— the warning notice and the recycling sign—offers a framework for the arguments made in this book. I originally intended *Strange Weather* to be a book about the cultural politics of technology. But in the course of writing I was increasingly confronted with ecological questions that undermined much of the familiar terrain of writing about technoculture, traditionally divided between emphases on liberatory technologies, on the one hand, and technologies of social control, on the other. For advocates, like myself, of a cultural politics that is often termed left libertarian, and that tends to see most "limits" as socially induced for the purpose of regulation, or even repression, the ecology movement's cardinal lessons about limitations are especially important to examine and consider. That is why I am quite sympathetic in this book to the caution advised by Murray Bookchin, who maintains that conditions of scarcity and deprivation recommended in the name of limits are always the result of social decisions not natural imperatives, and that the social freedom to be derived from recognizing limits can only be attained through choice and under conditions of postscarcity abundance.[1] Refusing any *direct* correspondence with geological or natural scarcity, Bookchin's version of social ecology insists that the imposition of "natural" limits is linked to systematic forms of social domination. This is a much more valuable perspective than that offered by cultural patricians like Christopher Lasch, who has recently employed arguments about the natural ecological limits of progress to recommend a return to the values of lower-middle-class culture, because those values are generally respectful of "limits" in all things and are morally organized around the family, the church, and the neighborhood community.[2] As Bookchin's radical humanism and Lasch's conservative moralism indicate, and as this book tries further to describe, the language of "limits" can have different meanings in different contexts, some very progressive, some not—just as the very language of "freedom" that Bookchin takes for granted can be used as an instrument of regulation as much as a discourse of liberation. "Limits," like "nature," or "the environment," is already one of the more important political keywords of our times; in the 1990s, we can expect each to carry more than its fair share of the burden of representation, and we must be prepared to account politically for the resulting variety of interpretations, each recommending this or that drastic course of action.

There does exist today a sophisticated cultural criticism that can discuss precisely these various meanings in their various contexts, but after twenty years of the modern ecology movement we have little that can properly be called a green cultural criticism. This book is intended, in part, to be a small contribution toward such a school—surely one of the more pressing challenges for cultural politics in this decade. One of the stories to be told by such a criticism will be about the history of a Western humanist culture devoted to the unrestrained development of human and social growth. The relation of that humanist history to the history of scientific and technological development is intimate and fundamental. Increasingly, we have come to see that C.P. Snow's famous intellectual divide between the "two cultures" is not so wide. While the specialization of academic knowledge has increased, and while what Snow, in 1959, called "the gulf of mutual incomprehension" between scientists and humanists has broadened, the two cultures are versions of the same story about development and growth. They flourish today side by side in academic and research institutions that are highly regulated by ties to the growth-driven logic of the corporate state. This is more evident in the evolved practical form of these cultures, the military–industrial–media complex, whose interlocking interests are increasingly well coordinated and increasingly difficult to bypass. This synergy was demonstrated as never before at the onset of hostilities in the Persian Gulf, when the precision coordination of military and media operations coalesced in an hour-by-hour presentation of the sublime spectacle of airborne pyrotechnics in the Gulf skies. The TV fireworks display of rockets exploding and smart bombs homing in on their targets was a warped tribute to the capacity of an advanced technological culture to organize its most exploitative interests in such a seamlessly ordered way. Here, in a war waged explicitly over neocolonial control of nonrenewable resources, the state and the media jointly offered a spectacular advertisement for another forty years of the permanent war economy, sustained by the uninterrupted flow of cheap fossil fuels. The subsequent management of the ecological consequences of the war—presented as the quirky result of "environmental terrorism" on the part of an Arab madman—glossed over the massive ecological threat posed daily by the multinational economy of militarism.

Even the cost–benefit analyses provided by the new corporate planetary managers show that our culture cannot afford a continued investment of scientific and technological capital in such an economy. In the years since Snow argued that scientists "have the future in their bones," while humanist intellectuals, his "natural Luddites," spent their time "wishing the future did not exist,"[3] ecologists have added the warning that the future may not exist after all.

Better living or quicker dying through chemistry? It is high time that cultural critics, often typecast as technophobes, played more of a techno-literate role in challenging the inevitability of both of these judgments: the technically sweet, theme-park future already germinating in the "bones" of science, and the dark, disastrous eco-future depicted in the survivalist scenarios. To mount that challenge and to contribute to alternative futures entails more than a schooling in technoliteracy; it also involves heeding a few historical lessons. There is a long tradition, especially in North America, of cultural critics who did not seek refuge in the Waldens or the "organic villages" that Snow saw as occupying the preindustrial fantasies of literary intellectuals in his day. This other intellectual tradition, whether bolstered by the strong native ideology of populist or progressive futurism, or by neo-Marxist beliefs in technology's liberatory potential, saw technology as an ally of progress and democracy; it is a tradition that has survived into the age of postindustrial high technology and the electronic utopia. As this book attempts to show, the history of that tradition charts the debasement of the idea of "technocracy," once the last word in rational intellectual leadership over society, but today the embodiment of soulless bureaucratic decision-making, abetted by the vast processing power of new information technologies. In this history the "future," traditionally seen as the utopian home of progressive left traditions, has become the favored environment of elite military and corporate planning and forecasting.

But there are no past "golden ages" here to dream into the future, and no simple technological solutions. The task of renewing left traditions of technofuturism today draws its appeal, for the most part, from the utopian underpinnings of the new social movements. Living differently in the future involves living with differences—of race, gender, nationality, sexual preference, and class—that did not feature in the imagination of radical technocracy in its heyday in the 1930s. Nor can we expect the high-density cyberpunk imagination of today's youth to be especially hospitable to preindustrial fantasies about the "natural life" that fed the counterculture's critique of technological rationality twenty years ago. The politics of nature today encompasses much more than pioneer visions of decentralized agrarian utopias or the atavistic appreciation of deep ecologists for uninhabited places. "Nature" is now also a battleground over the technobody: the politics of reproductive freedom, immune deficiency, biogenetics engineering, worker safety, police surveillance, narcotraffic, clean air, homelessness, militarism, and so on. In the thick of these battles with the corporate state over intimate, everyday environments, there is little to be gained from holding on to the traditional humanist faith in the sanctity of the unalienated, "natural" self, nobly protected from the invasive reach of modern science and technology. Nor can we always depend on the piratical appropriation, for countercultural uses, of technologies architecturally designed as instruments of control. Short-term survivalist actions may require such resourceful guerrilla tactics, but the longer-term task involves promoting the need for new technological designs, no longer in the service of bureaucratic rationality and controlled processing.

Strange Weather was conceived as an extension of some of the arguments in my last book *No Respect: Intellectuals and Popular Culture*, especially those that examined the cultural authority of intellectuals and the role they play in regulating taste cultures. My plan was to explore the reach of these arguments into the world of science and technology, which cultural critics have come to declare off-limits; or where, as assassins of objectivity, they are considered personae non gratae. As I lacked the training of a scientific intellectual and the accompanying faith, however vestigial or self-critical, in the certainties of the scientific method, it became clear all too quickly that my point of identification, even as a demystifier, could not be with official or "high" scientific culture. My position, then, became that of a cultural critic examining not only the power and authority of the claims made for science and technology (the dominant languages in our knowledge hierarchy), but also the various responses to these claims in the popular, lay, or public culture at large. While I occasionally analyze the language, philosophy, and rhetoric of the dominant scientific claims, my chief interest lies in describing how various scientific cultures—sublegitimate, alternative, marginal, or oppositional—both embody and contest these claims in their cultural activities and beliefs. Consequently, two related critiques operate here. The first is addressed to the ways in which technocratic elites—intellectual and political—have molded and regulated public opinion about the role of science and technology in shaping our future. The second is addressed to the different versions of and various challenges to these elite languages in popular and alternative cultures.

In pursuing the second of these critiques, I have devoted a good deal of attention not only to generic popular cultures like science fiction, but also to alternative cultures like New Age that are subordinate, marginal, or opposed to official scientific cultures governed by the logic of technocratic and corporate decision-making. This book is not, however, a celebration of the "resistant" qualities of these subordinate cultures. In this, I depart from what is taken to be an orthodox tenet of much of the recent work done in cultural studies. Far from being oppositional, the relationship between subordinate and dominant scientific cultures is as complex and interrelated as the relationship between popular and intellectual cultures in the realm of taste. Consequently I focus on how the authority of dominant scientific claims is respected and emulated even as it is contested by apprentices, amateurs, semi-legitimates, and outlaws who are detached in some degree from the authoritative institutions of science. While they are ostensibly opposed to the elitist, corporate workings of the official scientific and technical research environments, today's scientific countercultures share many of the methodological norms and claims about absolute truths in nature observed by establishment science. Indeed, some of the maverick, libertarian values espoused by countercultures run parallel with those prized in the entrepreneurial vanguard of corporate research and development. In this respect, the former play an experimental, and, occasionally, morally corrective, role for a dominant science culture that nonetheless deems their activities to be illegitimate and unscientific.

Most of the chapters in this book take as their backdrop the widespread influence of the information revolution—the chief capitalist revolution of our times, and the source of the new utopian communications order that high-tech evangelists have been proclaiming for almost thirty years now.

While there are many good reasons for doubting the gospel of silicon positivism, there are also lessons to be learned in modernizing our ideas about the workings of cultural power in a technocracy that claims to rule by knowledge, expertise, and smartness rather than by inherited capital, whether financial, cultural, or oligarchic. Earlier proponents of technocracy, in the 1920s and 1930s, challenged capitalism in the name of streamlined efficiency, promising a less wasteful, more rational system of economic and social life, with benefits more equitably distributed. Technocracy has long since been dealt a hand in the power structure of capitalism (which is increasingly dependent on science-based industry), while its efficiency logic has come to prevail over the scientific management of everyday life. The systematic effects of this social engineering have been widespread, from the reorganization of labor to the industrialization of culture and entertainment. Despite nominal appeals to rationality and progress, however, the gospel of the profit margin remains a more powerful doctrine than the gospel of efficiency. Capitalist reason, not technical reason, is still the order of the day.

In the progressive camp, technocracy's promise of a more rational society has been discredited, above all, for its reliance upon the forecasting of experts or technical elites, thereby excluding any provision for democratic decision-making procedures. Chiefly for this reason, technocracy is equated today with sluggish, bureaucratic planning that does not respond well to popular desires for a more creative and diverse, less standardized way of life. This mistrust of centralized decision-making has even carried over into popular attitudes about the more glamorous field of technology. The successes of postindustrial boosterism may have ushered in a new age of official technology-worship in the 1980s, but the revival of the faith has not met with such a broad or deep popular response as in earlier periods, such as the Machine Age of the 1920s and 1930s, for example, or, to a lesser extent, the Atomic Age of the postwar years. Skepticism about the social, economic, and environmental costs of technological development has become a permanent, often militant, feature of public consciousness. Cynicism about the remoteness of technocrats and fears about "technology out of control"—no matter how unfounded these fears are in their assumptions about the autonomy of technology—are both now regular conduits for the expression of ecological values. The mystique of technology continues to command admiration, especially in war, but decades of accumulated technoskepticism now make us wary of according to technology more power than it already has in our daily lives.

Alongside the discredited ethic of technocracy lies the growing legitimation crisis of scientific rationality itself. It is safe to say that many of the founding certitudes of modern science have been demolished. The positivism of science's experimental methods, its axiomatic self-referentiality, and its claim to demonstrate context-free truths in nature have all suffered from the relativist critique of objectivity. Historically minded critics have described natural science as a social development, occurring in a certain time and place; a view that is at odds with science's self-presentation as a universal calculus of nature's laws. Feminists have also revealed the parochial basis in the masculinist experience and ritual of science's "universal" procedures and goals. Ecologists have drawn attention to the environmental contexts that fall outside of the mechanistic purview of the scientific world-view. And anthropologists have exposed the ethnocentrism that divides Western science's unselfconscious pursuit of context-free *facts* from what it sees as the merely pseudoscientific *beliefs* of other cultures. The cumulative result of these critiques has been a significant erosion of scientific institutions' authority to proclaim and authenticate truth.

Critics like Stanley Aronowitz see science not as the realization of universal reason but simply as an ideology with a power that extends well beyond its own institutions, organizing every sphere of our daily lives.[4] In the face of such an awesome and pervasive presence, the task of challenging and demystifying the objectivity of science's claims has been paramount. But some critics fear that the disintegration of all fundamental claims about secure knowledge condemns science's dynamic social potential to the doldrums of relativism; where, in the absence of progressive winds, a paralyzing calm prevails. There is much to be said for Donna Haraway's caveat that "science has been utopian and visionary from the start: that is one reason 'we' need it."[5] We will always need sharp critiques of objectivity, because we will always need to be able to show that any picture of the world purporting to be "natural" and fundamental is in fact heavily underscored by particular moral and political beliefs about nature and social behavior. That, after all, is how we recognize the effects of *difference* at the heart of a rationality that necessarily ignores these differences in its pursuit of universal judgments. But there is also a critical responsibility to renew the demand for fresh futures that will be more radically democratic than our own present. We therefore need a scientific culture that can learn from differences of class, gender, race, and biology, and that can transform notions like progress and objectivity in order to address these differences and the social inequalities created in their name.[6] That kind of culture will have to be actively constructed from embodied scientific knowledge about the environments we inhabit, rather than from the kind of knowledge, traditionally espoused by science, that empirically separates the environments from our lived experience of them.

It is beyond the scope of this book to detail the features of such a scientific culture, but since it will be a fully material "culture," not merely a set of cold equations, I hope that the cultural issues and questions I do address might contribute to the debate about its future. If that is why this book insists on seeing science as culture, the reader ought to be warned that the result, in these pages, is a good deal of strange weather. Culture, after all, is supposed to be that which is unscientific, that which despoils the golden rule: the contagions of rhetoric, the distortions of passion, the subversions of imagination, and the obduracies of ritual, faith, folklore, and convention. Culture is supposed to be about the molten liquidity of experience, whereas science aspires to the solid dimensionality of fact. But the empirical naming and knowing of the physical world is nothing if not a culturally expressive act with fully political meanings. Our current ecological crisis demonstrates that the consequences of such artificial divisions of knowledge about the world have been quite deadly, and that is why the crisis of science today is also a crisis of our culture—guided, as it has been, by assumptions about its unlimited access to resources, both rational and natural.

A brief word about method. The arguments I present in this book about the shape and the claims of scientific (counter)cultures and communities are not the result of the kind of close ethnographic study that has become such an exciting feature of recent work in cultural studies. My work does rest upon a fairly extensive study of the ideas and claims of such communities, and it comes out of a familiar, sometimes personal (sometimes not), working knowledge of the ways in which they act and think. But it does not aim at an exhaustive "thick description" of these communities. What I try to do is provide a polemical sketch of the salient activities and claims of the respective cultures in a way that invites general, rather than exclusively scholarly, debate. In pursuing this aim, I have tried to balance the risk of inevitably misrepresenting the lived experience of some of these communities against the decision to make more general points about how these activities are represented and discussed, again in a more public setting. In the same vein, I chose objects of study from across the cultural spectrum, primarily because they have become topics of interest widely debated in public media: New Age, hacking, cyberpunk, environmental decay, futurology, global warming, and others. While I have tried to preserve a linear form in the presentation of my material, building on arguments and histories from chapter to chapter, each of these studies might also be read as a self-contained discussion of the questions raised therein.

The result, I hope, sounds more like a reasoned public discussion of issues concerning science and technology than the conclusive findings of a specialist in the field. If the technological future is an issue that concerns us all, then decisions about its shape cannot be confined to the exclusive domain of specialists and experts. Nor can we ignore the personal passions that must be brought to bear upon such decisions. It is from the crucible of these passions that this book takes its title, because the weather serves as one of my own ways of making sense of the natural world's role in my psychosocial life. For me, the weather has always been the most reliable witness that the world would soon change. If I had to come up with a metaphor for my kind of futurology then I imagine it would be something like *social meteorology*. A meteorology on the ground and of the ground, as well as from the air, the traditional vantage-point for forecasting. A meteorology that builds fronts from below in addition to one that sees the fronts coming in advance. And, above all, a meteorology that can explain some of the desired connections between social life, natural life, and economic life.

CHAPTER ONE

NEW AGE — A KINDER, GENTLER SCIENCE?

I now have attained freedom just as fully and really as a runaway slave might have in the pre-Civil War period.
RESPONDENT TO A NEW AGE QUESTIONNAIRE

In the fall of 1989, I attended my first Whole Life Expo for Body Mind Spirit in New York City. One of the New Age community's primary trade conventions, the Expo featured a bewildering array of booths, exhibits, lectures, panels and workshops. The whole spectrum of New Age interests and practices was represented: from the sober company of elder statesmen of the Rosicrucian faith, regaling the modern mind with ancient mysteries, to the glamorous presence of the much-hyped mind machines, the latest high-tech gadgets in the field of "brain expansion." Ranged somewhere between these extremes was the usual eclectic and colorful mix of healers, psychics, holistic foodsters, folk metaphysicians, psychotechnologists, UFO contactees, crystal therapists, dolphin advocates, channelers, and a roster of New Age stars such as Timothy Leary, Marilyn Ferguson, Lynn Andrews, Elizabeth Kubler-Ross, Uri Geller, Kevin Ryerson, Whitley Strieber, Sun Bear, Wallace Black Elk, Gary Zukav, and Deepak Chopra.

One of the sessions I attended was designated as the "science panel," visibly sponsored by a company called ELF Cocoon International that specializes in subtle energy instruments. The speakers on the panel were addressing the topic of Extra Low Frequency (ELF) electromagnetic radiation, in response to a public debate intensified by recent revelations about the acute health disorders caused by 60Hz fields generated around low-voltage power lines, computer VDTs, high-tech home appliances and the like. Scientists have long believed that non-ionizing EM energy had no adverse biological effects other than heating human tissue, but strong evidence now links low-frequency radiation to a host of cancerous conditions as a result of daily exposure in the home and workplace.[1] While the paramount danger to public health of these EM fields has been suppressed by a combination of military and industrial interests, this topic has long been a paranoid favorite among right-wing conspiracy theorists, concerned about the clear Soviet lead in psychotronic warfare, and all too eager to trot out the latest allegations about the use of giant Tesla magnifying transmitters to aim over-the-horizon radiation at locations like the West Coast of the US.

The make-up of the panel itself had been skillfully arranged. The first speaker was a smooth, nattily dressed publicist of popular technology, who put the case for responsible and ethical development of ELF technology. He was followed by a congenial scientist who outlined some of the features of the debate about environmental electronic pollution: its effects upon new and unexplained diseases, and the possibility of electromagnetic erosion of our immune systems. Next up was the star, Robert Beck, virtuoso inventor of the stroboscope and other electromedical stimulators, who tempered his warning about potential government use of ELF to control the emotional lives of mass populations with an account of his own scientific discovery in the 1960s that the brainwaves of healers and mystics shared the same frequency—7.83Hz—as the earth itself, vibrating in the ionosphere. Also on the panel was the obligatory Soviet scientist, soberly testifying to the advances made by the Soviet scientific community in measuring the parapsychological aspects of ELF radiation. The most volatile presence, however, was a crew-cut nerdy scientist, straight from central casting, who delivered a loud diatribe (negative discourses are few and far between at New Age gatherings) against the unchecked entrepreneurial development of bioenergetics "wireheading" devices for health and human potential purposes. Referring to the dangers involved in this unregulated practice, he chose, as an example, the perilous acceptance of such technologies by sectors of the Aids community as a way of treating HIV infection, and concluded by warning that the government would use any of the "bad" examples it could find to crack down hard on a bioenergetics field that was struggling for legitimacy. Similar arguments have been made by scientists who claim that activists' adverse publicity about the dangers of low-frequency radiation would punitively affect the funding of research into the electrochemical workings of the brain. As in so many public debates of this nature, scientists' responses are often determined by the limitations on their ability to maintain funding for their own specialized research. A final note of conciliation was introduced by the panel moderator, who reaffirmed that her company, ELF Cocoon International, was not in the business of making snake-oil technologies, that their products were backed by hard science, but that we still had to watch out for the "bad guy" frequencies that were hiding in there among the "good guy" frequencies—a distinction, it was implied, that was not always observed by her company's competitors in the field.

What we heard on this panel, then, were at least three forms of address—a hard science perspective, an appeal to the interests of the holistic community, and the voice of persuasive entrepreneurship. Carefully selected not only to represent "science" to a New Age audience but also to cast suspicion on other competitors in the business of selling bioenergetics products, this Expo panel ensured that the dissenting voice that came closest in style and polemic to invoking the authority of hard science principles against the development of such products was one that could ultimately be assimilated within the desired agenda of the panel's commercial sponsors, whose products are intended for use by a community that considers itself to be the home of a truly radical alternative to the authority enjoyed by scientific positivism's high priests. While the positivist orthodoxy of establishment science is ritually vilified by New Agers as elitist, left-brainist, and inhumane, it would be a mistake to assume that the spirit or the name of science is persona non grata in New Age circles. As Stanley Aronowitz has argued, the preeminence of science in our knowledge hierarchy is echoed and emulated, to some extent, by all cultures subordinate to it, even when their identity is marked as alternative.[2] Consequently, assuming the mantle of a rationalist style is an indispensable discourse for those whose business it is to contest orthodox claims about the natural world.

Despite his professionalist conviction, the lone dissenting voice on that panel of bioenergetics advocates was speaking from a position highly marginal to the licensed rationalist center. His field was still struggling for legitimacy, and so his criticisms reflected the concerns of an entire discipline anxious to distance itself from anything that would be seen as "far out." The more official and centered voice of condemnation against the New Age community can be found in what are often characterized as the witch-hunting activities of CSICOP (the Committee for the Scientific Investigation of Claims of the Paranormal). CSICOP is an international "inquisition" of mostly academic ghostbusters, set up in the mid-seventies not only to combat the rise of the Christian fundamentalists' creationist claims, but also actively to police the boundary between science and pseudoscience contested by a host of New Age alternatives to institutional scientific orthodoxies.[3] While most of CSICOP's polemics take the rather dull and predictable form of positivist debunking, some of its more popular spokesmen, like Carl Sagan, often advance the claim that the orthodox rationalist view of the natural world contains quite enough miracles—black holes, for example—to preempt our seeking out thrills on the murky or occult fringes of reason. Likewise, CSICOP's most successful vernacular strategist is the Amazing Randi, a professional magician who claims to replicate exactly, by sleight of hand, the feats and miracles achieved by practitioners of paranormal behavior like Uri Geller. Rationalists, it is thereby suggested, are a superior breed even when they masquerade as professional non-rationalists. But such exposés of paranormal activities, whether in dry polemics or in showbiz, are always conducted through appeals to the kind of experimental certification that rationalist science has established as the single standard of truth and reason in our dealings with the natural world. In this respect, they might be seen as affirmations of faith in the world-view of a particular culture, which cannot afford to recognize the principles that underlie other explanations of the natural world. The existence of a ghostbusting organization like CSICOP is as much a symptom of the crisis in scientific rationality and materialism as it is a grudging acknowledgment of New Age's resurgent interest in the minority non-rationalist traditions of Euro-American culture.

Needless to say, many of the audience at this Expo panel, including myself, flocked to the trade booths afterwards for a free joyride on the wild brain machines, about which I will say more shortly. For the moment, I shall venture that what was going on at this panel presentation, seemingly far from the madding crowd of channelers and mystics speaking in tongues in neighboring rooms at the convention, could be seen as a kind of composite *explanation* of the kind of "rationality" that the New Age movement, however inchoate, has come to present as an alternative, though not necessarily in opposition, to the dominant paradigms of scientific and technological rationality.

First and foremost, the panel's existence assumed its audience's prior interest in the effects of low-frequency energy. Their practices, as holistic healers for the most part, are grounded in various philosophies of "energy," whether vitalist or astral. For all the sobriety of the panel's proceedings, there was no great gulf between its discussants' claims for the legitimacy of bioenergetics research and the channeling going on next door. ELF, the topic of the panel, is held to be the explanatory basis of many parapsychological and paranormal phenomena. It is the frequency claimed for the multifarious properties of crystals, the frequency at which psychic healers, mystics, extraterrestrial contactees, channelers, and other psychotechnologists are said to operate, and the preferred frequency of the various cosmic energies (*ch'i*, *prana*, *mana*, *kundalini*, odic energy, orgone energy, aura, and others) that are the foundational components of so many New Age metaphysical systems.

But the panel's organizers also knew a thing or two about their audience's love–hate affair with rationalist science. The fierce independence of the New Age community, more visible perhaps in the activities of the neighboring channelers, is increasingly compromised by its evangelical desire to move beyond a marginal status in the legitimate scientific world. Consequently, the claim to an *alternative* world-view, distinct from orthodox rationalism, increasingly rests upon arguments that this claim is nonetheless unjustly *excluded* from or suppressed by the dominant scientific paradigm. A certain degree of common cause is thus established with those within the legitimate scientific community whose work is locally contestatory and thus marginalized or suppressed—in this case, the activist researchers who have been challenging established wisdom about the biological effects of EM radiation. The commonality of interests often rests upon a self-presentation as victims of coordinated neglect, even persecution, by powerful vested interests. Holistic practitioners, interested in exploring the liberatory "human potential" of bioenergetics, may have different motives for their interest in this research than do activist scientists; but their respective interests are commonly served by drawing public attention to the debate about low-frequency energy. Ultimately, some part of the holists' desire lies in their hope that rationalist science, no matter how fundamentally impaired by its materialist premises, will be able to prove the legitimacy of New Age claims about the electro-vitalist basis of human life. As a result, the shape and language of holists' claims about alternative scientific knowledge are mediated through appeals to the rationalist language and experimental procedures of the dominant paradigm. Contestation, as I have suggested, would be impossible otherwise. But it is in contexts of this sort, at the most rationalist end of the New Age spectrum, that we can see the tension within a social movement founded on an *alternative* scientific culture, distinct from dominant values, that is increasingly obliged to lodge *oppositional* claims lucidly obedient to the language and terms set by the legitimate culture. Therein lies a story about the contradictions of New Age culture as it exists today, but it is perhaps also a story about the evolutionary structure of all such social movements that exhibit a slow development from marginal, visionary origins to mainstream encounters with professionalization and institutionalization.

Perhaps there is nothing more to be observed here than the old Marxist lesson that a dominant culture has the power to engender an opposition in its own image. I would have to say that it is a good introductory lesson, quite applicable to many of the features of the New Age movement as they exist today. By New Age, I am referring to that exotic subculture whose cultural practices and beliefs have attracted tens of millions in the West—and increasingly large numbers in the Soviet Union, where alternative, metaphysical voices have begun to appear daily on television screens in a culture that has made a state religion of science and technology for the collective good. It is a useful lesson to bear in mind if we want to trace the swollen flood of philosophies and practices with a rich and diverse historical heritage that emerged in a relatively coherent form in the wake of the 1960s counterculture, and underwent a sea-change as they developed their own institutions and networks in the 1970s and 1980s, often within an entrepreneurial milieu lovesick with the romance of the technological sublime. But there are other, more complex lessons that I wish to draw from the widespread influence of New Age ideas in our culture; lessons, for example, about the crisis of science and technologism; the crisis of "nature"; and the crisis of materialist individualism. While most New Age practices today are still seen as restricted to a minority culture, the influence of their ethical principles is quite mainstream and middle class, permeating suburban life and corporate philosophy alike. One need only tune into popular TV talk shows like the *Oprah Winfrey Show* to hear the language of "growth" and "potential" in full flow. To understand the logic of New Age's language of individualism is crucial for anyone who wants to understand the ideological shape of North American culture today.

New Age individualism, with its overriding appeals to personal growth, draws upon a long history of decentralized minority traditions in Western culture: for example, the Christian esoteric tradition, with its various Gnostic sects and heresies; or the Transcendentalist, Spiritualist and self-help movements of the last two centuries, each preserved in noninstitutional ways in avant-garde or bohemian circles (or jealously guarded by masonic and aristocratic cliques), while occasionally exercising a broad political appeal to larger constituencies in revolutionary or reformist historical moments. For many centuries, those traditions of radical individualism have been influenced by the more established Oriental social philosophies of the Chinese and Ayurvedic medical traditions. All of these influences, grounded in naturalistic, holistic, and pantheistic sciences, have been systematically displaced and repressed by Western empiricism.

Notwithstanding these important histories, however, any explanation of the currency of the social movement based on New Age individualism must address the more recent socio-historical conditions in which appeals to "personal growth" have taken root. In this respect, we might consider the New Age discourse about "growth" as a response to widespread anxieties about the official Western ideology of growth and development that have been generated, primarily by ecolological concerns, in recent years. In the pages that follow, my aim is to draw some lessons from the study of alternative cultures like the New Age movement, and to show how the "solutions" and new wisdoms about "limits to growth" offered by such a culture at once contest and reinforce dominant values about human and technological growth and development.

New Age rationality, as I briefly describe it here, can be seen as a countercultural formation in an age of technocratic crisis. This crisis appears at a time when the official legitimacy accorded to technology-worship has guaranteed it the status of a new civil religion in North America, perhaps the only possible millenialist home that remains for offical versions of the emptied-out American dream. Yet it is also a time when faith in modern science's founding sacraments—its claims to unimpeachable objectivity, axiomatic certitude, and autonomy from the prejudices of power—is rapidly disintegrating under the pressure not only of demythologizing critics and activists within the priesthood, but also from the thoroughgoing historical critiques of scientism waged by feminists and ecologists with one foot in the door, and from public disaffection with science's starring role in the grisly drama of global degradation.

Given this state of affairs, it is hardly surprising to find some indications of this crisis in the platform of New Age "rationality": on the one hand, obliged to emulate the dominant, rationalist discourse, and on the other, condemned to the fringes of pseudoscience and the twilight zone of reason by the guardians of orthodoxy. The ultimate purpose of the Expo panel, for example, was to establish safe passage for its sponsor's high-tech products, whose responsible development, in the face of the known dangers, is tied into assumptions about the perfectibility of human potential and growth—an aim that is quite compatible with even the most idealistic goals of scientism. On the other hand, the stories told by some of the panelists about thwarted research, government conspiracies, scientific elitism, environmental abuse, and runaway entrepreneurial logic reveal the bottomless paranoia generated by the clash of the hard science-oriented fringe of the New Age spectrum with the decimated core of objectivity that haunts the ethical pursuit of science today. By contrast, the larger and more humanistic sectors of the New Age community have made common cause with quantum physics, finding among the more speculative adherents of that discipline a tolerance for mysticism that complements their own holistic metaphysics, and a new *raison d'être* for closing the gap between the "two cultures."

Again, it is no surprise that paranoia—a political symptom, in this case, of much more than the *ressentiment* of the excluded—should run deep in the more empirical contests waged by New Age over the troubled body of scientific objectivity. While it is true that the contest over legitimacy at the most empiricist end of the New Age spectrum is more *articulate* than among those who lean upon the metaphysical critiques of scientific rationality, it is probably not different in kind from these more occult contests, which share a world-view as logically consistent in its own way as the rationalist explanation of the natural world offered by modern science. Holists who subscribe to a cosmology of dynamic forces and energies may not share the mechanistic world-picture of separable causes and effects, but they do share an everyday world in which rationalist assumptions are dominant; therefore, any attempt to promulgate their claims must make its persuasive mark in a culture saturated with these assumptions.

THE DEMARCATION DEBATE

A long history lies behind the anxious intimacy of New Age metaphysics with scientific thought. Practitioners of the occult who subscribed to the supposedly ancient Hermetic tradition, and who were hit hard by the Reformation's anti-supernaturalism, ultimately leaped to embrace science as one of their primary vehicles for a critique of orthodox Christianity. For centuries, this embrace of science helped the heretic traditions to survive. Many of the health and parapsychological cults of today are the direct descendants of the heretical, science-loving, metaphysical movements of the eighteenth and nineteenth centuries, like Magnetism, Mesmerism, Spiritualism, Swedenborgism, and Theosophy, or the more unapologetic medical sects of Phrenology, Hydropathy, Vitalism, Chronothermalism and other non-allopathic disciplines. Not only were these practices grounded to some degree in the principles of empirical science and inductive reasoning, they were also caught up in the reform movements of their day, consequently appealing to a broad and popular social constituency. Aside from this appeal to popular forms of politics, philosophy, and spirituality, their medical claims, in retrospect, stood up well enough to some of the barbaric prescriptions offered by the orthodox medicine of the day, especially in the age of Heroic Medicine, when physicians employed violent remedies such as purging, bleeding and the use of calomel (mercurous chloride). By the time of the Flexner Report in 1910, the American Medical Association had succeeded in outlawing many of the pseudoscientific practices, but not before the medical establishment had incorporated some of the lessons of the alternative, reformist and healing practices, and not before some of its early redbaiters had sounded a link between the "pseudoscientific" trends and the specters of socialism and feminism. As the editors of the *Boston Medical and Surgical Journal* complained, the medical practitioners who were currently "running after this hydropathic mummery" were last year "full of transcendentalism, the year before of homeopathy, the years before of animal magnetism, Grahamism, phrenology. Next year, they will be Fourierites, communists, George Sandists, etc."[4]

A similar kind of story might be told about the role of the alchemical tradition at the time of the formation of modern science. Inheritor of the Hermetic quest for a universal specific—a magic elixir, powder, potion, or touchstone that prolonged life and cured all ills—alchemy was *the* medieval science of matter, suffused with the ideology of animism, but very much in cahoots with the industrial logic of mining, dyeing, glass manufacture, and medicinal preparation. The displacement of animism by mechanism, and the subscription of new artisanal groups to laissez-faire market principles that openly challenged the secretive trade knowledge of the older craft guild system, sounded the death-knell of the alchemists' power. Alchemy was increasingly cast as heretical by a Reformation dressed up in its rationalist Sunday best, and survived only in obscurantist groups or in politically radical circles like the leveling sects of the English Commonwealth.[5] As Morris Berman has argued, however, it would be a mistake to think that the mechanistic world-view of the new science simply excluded or excommunicated the Hermetic tradition (in fact, it was driven underground) as illegitimate, demonizing it as a cover for quacksalving charlatans. Rather, the foundational components of magic—the capacity to alter, transform, and thereby dominate nature—became central, albeit in a distilled form, to a mechanistic world-view bent on systematically governing the natural world.[6] Alchemy, after all, was nothing more than a method for transmuting nature into energy, and few could deny that this has become one of the technological projects, with a vengeance, of modern science. The alchemists' dreamy pursuit of Nature's truth was no less shared by Galileo and Newton (a closet occultist himself). By Newton's time, mechanistic science had recycled many elements of the magician's enterprise and relocated the early research laboratory milieu of his kitchen to larger institutional spaces, equally in the secular service of earthly powers. As the governing premises of animism were displaced by the new materialism, the science/pseudoscience border, ever shifting, was accordingly redefined. The esoteric sciences, now subordinate rather than dominant, were obliged to incorporate a shell of inductive reasoning in order to wage their claims, just as they would learn to survive under cover of rationalism's long and unholy war against the established churches.

The sound and fury generated today not only by the flourishing of holistic health and occult metaphysics (often self-styled as the new alchemies), but also by the revived fundamentalist agenda of creationism, demonstrates that the demarcation debate about "false" and "true" science is not one whose significance lies only in some parochial past, safely relegated to a period of science's infancy. On the contrary, histories of the sort that I have sketched only briefly show that any such demarcation is *always* historically specific, determined by the cultural and ideological circumstances of its day, and thus by the particular claims that "science" and "scientists" make for themselves in a particular time and place. Moreover, the borders between scientific cultures are not only semipermeable, they are rigorously patrolled in the interests of powerful institutions; increasingly so, since the modern laboratory itself became a model of power, and since corporate capitalism started to gamble its future on the magical elixir of basic patents.

In this respect, it is perhaps worth drawing an analogy between the demarcation lines in science and the borders between hierarchical taste cultures—high, middlebrow, and popular—that cultural critics and other experts involved in the business of culture have long had the vocational function of supervising. In both cases, we find the same need for experts to police the borders with their criteria of inclusion and exclusion. In the wake of Karl Popper's influential work, for example, falsifiability is often put forward as a criterion for evaluating scientific authenticity, and thus for distinguishing between the truly scientific and the pseudoscientific. But such a yardstick is no more objectively adequate and no less mythical a criterion than appeals to, say, aesthetic complexity have been in the history of cultural criticism. Falsifiability is a self-referential concept in science, inasmuch as it appeals to those normative codes of science that favor objective authentification of evidence by a supposedly dispassionate observer. In the same way, "aesthetic complexity" only makes sense as a criterion of demarcation inasmuch as it refers to assumptions about the supposed objectivity of categories like the "aesthetic," refereed by institutionally accredited judges of taste.

I do not want to insist on a literal interpretation of this analogy, but it is an analogy that informs my own thinking as a cultural critic about some of the points I want to make in this chapter. A more exhaustive treatment would take account of the local, qualifying differences between the realm of cultural taste and that of science, but it would run up, finally, against the stand-off between the empiricist's claim that non-context-dependent beliefs exist and that they can be true, and the culturalist's claim that beliefs are only socially accepted as true. Ultimately, the power of science rests upon making and maintaining that distinction, and we ought to recognize that science's anxiety about authenticating its belief in truths is, in the truly Foucauldian sense, a question of power. Consequently, it is not such a great leap from seeing that categories of taste are also categories of power employed to exclude the unwanted to seeing that the power of scientific ideology rests upon its unwillingness to question the role of the powerful institutions or sponsors whose interests are not only heavily mortgaged in the demarcation debate, but who are also well served by the hireling scientists who referee it. Cultural critics have, for some time now, been faced with the task of exposing similar vested institutional interests in the debates about class, gender, race, and sexual preference that touch upon the demarcations between taste cultures, and I see no ultimate reason for us to abandon our hard-earned skepticism when we confront science.

For my purposes here, the analogy with taste cultures helps make sense of certain "middlebrow" elements of New Age philosophy and culture when compared with the "highbrow" status of the official scientific cultures and the lowbrow gadget-fetishism of popular science. For the components of New Age culture are both middlebrow and alternative, a composite that can probably be found in any counterculture that has found some kind of breathing space, however marginal, within the dominant culture. On the one hand, the devotion to alternative, non-rationalist belief systems places New Age thought outside the hierarchical structure of cultural capital observed by the legitimate scientific culture. On the other hand, the New Age commitment to transforming science into a more humanistic and holistic enterprise involves taking on, to some degree, the structure of deference to authority that governs the institutional system of rationalist cultural capital. As a middlebrow scientific culture, New Age wants to be fiercely self-determining, but the path to establishing that authority leads through the obstacle course of accreditation that underpins scientific authority and marks non-institutionalized opinions as illegitimate.

If the components of New Age culture are a complex fusion of the alternative and the middlebrow, then critical precedents for describing such a culture are not so easy to find. The middlebrow, or what used to be so termed, remains one of the largest challenges for cultural studies, a field in which so many have leapt with alacrity to the analysis of popular cultures, uncovering elements of "resistance" that are comfortably remote from the levels of cultural capital more appropriate to the critiques of intellectuals. By contrast, the more contiguous field of middlebrow culture, when it is not passionately denigrated, is thought to be stale, flat and unprofitable, its politics unremarkable, and its pretensions devoid of the edge that comes with the everyday alienation, as at least one version of populist cultural studies would put it, endured by the popular classes.[7] I know, for example, that many of my fellow intellectuals think of New Age as the lowest of the low, and cringe on contact with evidence of its influence on their daily environments, while these same people can be passionately devoted to analyzing and talking about the cultural politics of TV soap opera. I do not want to distance myself from this habit, for I am as much a "victim," if you like, of this essentially intellectual tendency. It is a symptom of the logic of cultural capital that the culturally wealthy can afford the kind of downward mobility which sanctions their devotion to the popular, while they police the cultural order by deriding the sublegitimate middlebrow as only they can. The challenge for intellectuals, clearly, is to help create a more democratic cultural politics that would not be hamstrung by this logic.

Like most fellow intellectuals I know who grew up in the heady wake of the 1960s counterculture, my experience with holistic friends, acquaintances—and also, in my case, an ex-spouse with New Age devotions—has been touched by a shared passion for the utopian, experimental cultures that survive to some degree in the New Age community. However, the history of our political and intellectual training has made us "outsiders," for the most part, in relation to this community, with whom we might share a commitment to forging social and cultural alternatives, and with whom we often share the field of grassroots activism. It should be clear, for example, from the marks of tone, style, and critical distance, that I am speaking neither from the New Age community nor on its behalf. But neither am I primarily interested in speaking against the New Age belief systems and their political implications. My position is more that of a speculative critic who thinks and feels that there are political lessons to be drawn from the shape and development of New Age culture. Consequently, a large part of my interest here lies in discussing not only the self-coherent logic that governs New Age (however eclectic its cultural range) as a movement, but also the major lines of conflict and contestation that have developed around the New Age community's challenge to established institutions of science and religion.

The disadvantage of this polemical position is that it neglects the more exhaustive, or deep, "ethnographic" study of cultural communities that has produced some of the most exciting developments in recent cultural studies. Although my research has involved little in the way of "field work," it does draw upon an extensive immersion in the literatures produced by and around the New Age community: a broad spectrum from the established book classics to the more eclectic flood of magazines, pamphlets and newsletters that has poured daily into my mailbox over the last few years. The traditional function of the polemical critic lies in making *interventions*, in using his or her position as an intellectual to enter into the more general public debate about the shape of cultural politics. There are many reasons to be critical of this tradition: it has reproduced elitist access to media and intellectual opinion; it has often fallen prey to the voice and style that signify a remoteness from the lived experience of culture; it has renounced the official authority that comes with exhaustive academic scholarship, just as surely as it has lacked the organic authority derived from involvement with active cultural communities. If public intellectuals today are learning these lessons in self-criticism, then the tradition of cultural studies is one of our best resources, for its strengths lie in learning how to respect the lived experiences of cultures other than the intellectual, and to draw out, rather than preach about or against, their political implications. But if the practice of cultural studies is to preserve its activist direction, then it cannot afford to give up a public voice that goes beyond the relativism of respectfully recognizing and appreciating all cultural differences equally. The problem with such relativism is that it tends to lose sight of the way in which cultural differences are themselves always a function of power; these differences, in other words, are never equivalent, and always unequal. The game of assessing the differences between cultures is not just a game of words, in which the stronger rhetoric wins; a great deal of material power is exercised through the existence of these unequal differences. Consequently, we cannot afford to relinquish the authority, derived from institutions and professions, that can be put to good use in contesting the official policing of cultural politics.

In science, probably more than in "humanist" culture, there remains the challenge of providing, as Donna Haraway puts it, "better accounts of the world," that will be publicly answerable and of some service to progressive interests.[8] To keep that obligation in mind is to resist many critical temptations in describing scientific cultures: the temptation to explain scientific claims as merely epiphenomenal effects of vested material and institutional interests or alternatively to try to separate the wheat from the chaff by isolating particular scientific claims, if any, that can be said to be truthful and thus immune to the contagions of power; the temptation to collapse all competing knowledge claims into a relativistic picture of warring subcultures or alternatively to romanticize those subcultures that are deemed "illegitimate" as havens of resistance. If the rallying-cry for a "science for the people" is still to stand for something that resembles an objective vision of the social good, then it depends on salvaging workable strategies from the vertiginous relativism that often results from extreme culturalist analyses of science's day-to-day workings. Such strategies, which will need to be locally relevant and not generally self-affirming, can no longer afford to appeal simply to optimistic or progressive versions of technocratic radicalism, in which a socially minded elite leads the way forward. They must also be addressed to the desire for personal responsibility and control that will allow nonexperts to make sense of the role of science in their everyday dealings with the social and physical world.

I do not believe that New Age culture has produced anything like a more consistently accurate account of the world than rationalist science. Rather, in its own eclectic and self-fashioned way, New Age has assumed a virtuoso, experimental role in reconstructing a humanistic *personality* for science— science with a human face. A kinder, gentler science. This appeal to personalism is deeply rooted in popular distrust of authority and the desire for self-control; it cannot be dismissed as a "petty-bourgeois" obsession. But the need to acknowledge the personalist appeal coexists with the need properly to socialize this concept of individualism, to show how self-responsibility can only be achieved by transforming social institutions that govern our identity in the natural world. In its embattled attempts to practice a science pirated and reappropriated from the experts, the New Age community feeds off the popular desire for more democratic control of information and resources. Because of its embattled position, however, the New Age community is also driven into defending the moral purity of an *alternative* scientific culture, which draws upon philosophical traditions, whether Oriental or archaic, that are not part of our socialized landscape in the West, and that consequently have much less purchase on people's commonplace thoughts and desires. All too often, the result is to dream us back to the prescientific and to the alchemists' kitchen—which, for all their charms, are rather claustrophobic places for us to be in the 1990s.

DIGITAL ALCHEMY

I have argued that non-legitimate scientific cultures, no matter how esoteric, unavoidably bear the impress of the dominant orthodoxy's language; this was as true of the nineteenth-century "pseudosciences" as of the alchemical traditions that played a role in the formation of modern science. Accordingly, the prevailing ideologies of high-tech systems theory and cybernetics provide not only much of the language and the conceptual logic, but also a good deal of the holistic framework (a result of their shared emphasis on the interconnectedness of systems) for alternative scientific cultures today. To illustrate this point, we might consider more closely the place within the human potential movements of these wacky brain-machine technologies that I earlier mentioned.

The more visible and commercially successful of these technologies (most are comprised of a digital programming device, headsets and an optical apparatus) have names like Synchro-Energizer, the Alpha Stim, RelaxPak, Graham Potentializer, Hemi-Synch, Tranquilite, Somatron, the Biofield Analyzer, Theta One, NeuroSearch 24, the Mindeye, Endomax, the Mind Mirror, Isis Surge Resonator, and the Alpha Pacer. They stimulate the brain through a coordination of flashing lights and strobes, electrical impulses, and magnetic fields directed at the cerebellum and hypothalamus; and utilize sound patterns and ultrasonic spoken words aimed at transcutaneous electrical nerve stimulation (TENS). Some employ biofeedback—an electronic system that monitors and informs the subject about the level of activity of selected physiological processes, feeding back tones or visual representations of bodily activity, information that allows the subject consciously to "control" autonomic nervous processes. For the most part, these gadgets are all adaptations of a medically approved "black box" used on arthritic patients to alleviate chronic pain, and also on epileptics. Official funding tends to go towards the development of "black box" use for medical science. New Age mavericks interested in using the machines for their personal growth have been obliged to develop them as entertainment products in the commercial market, where Congress is increasingly likely to crack down on advertised claims about their behavior-changing powers.

All these machines are designed to balance the left and right brain in holistic synchrony. In altered states of consciousness, brainwaves become slower and deeper, moving from beta in normal waking life (13–38 Hz), to alpha (8–13 Hz), to theta (4–7Hz), reaching delta wave levels in deep sleep. The operating hunch of these technologies is that the process of causation can be reversed; by first changing the brainwave frequency, one can induce altered states of consciousness. Most of the philosophical interest in this question derives from the brave new world of neuroscience, whose researchers claim that up to 90 per cent of the brain's neurons go unused in cognitive activities, and that brain cells, brain memory, and size of cortex are by no means destined to entropic decay, but can be expanded through the kind of external stimulation provided by the brain machines. As one enthusiast puts it: "It's as if we've all been given superbly engineered sports cars in which we've been putt-putting about without ever shifting out of first gear, never realizing that there were higher gears."[9] In fact, this kind of high-tech performance metaphor is quite typical in descriptions of the after-effects of a workout on a brain-booster: "Each time I emerged . . . I felt fine-tuned—like a superb motorcycle humming along a smooth highway at about 95mph ticking like a clock."[10] And even more masculinist yet: "My ninety-seven-pound weakling of a brain could pump mental iron and emerge as a slab-muscled two-hundred-pound heavy thinker capable of kicking sand in the face of any bully metaphysician on the beach."[11]

Other testimonies are more gender neutral, while promising a particularly vague kind of success in the world. An ad for Inner Quest technologies runs thus: "As Alpha and Theta brain waves are stimulated, you begin to totally relax, allowing you to absorb new and positive images and ideas. Goals become attainable. Attitudes become positive. Life becomes enjoyable." For the record, my own response to my first brain-boosting session was to fall fast asleep, a lapse to which I am prone anyway in the early afternoon. My next sponsor was more commercially upfront, and asked me what kind of narcotic high I wanted to simulate. I opted for a speed program, and for a little while at least, seemed to feel a tad more perky than usual.

Brain gyms and spas housing these mental Nautilus machines have sprung up all over California, as the high-tech ashrams of the physical fitness regime make room for the new mental fitness workout. Seductive claims of increased long-term and serial memory and intelligence are quite consonant with the Yuppie work ethic—more mental work, performed more efficiently than ever before. In the culture at large, the same code of efficiency extends to the realm of leisure time, where we have become accustomed to hearing the Yuppie idiom of "working" on a tan, or a salad. As a testimonial in one brain machine ad puts it: "it was like taking a two-week vacation in 28 minutes." The same ad establishes a demographic appeal that is quite specific: "Listen: I did everything in the 60s. I did all the retreats in the 70s. I've sampled most of the brain tune-up machines in the 80s."

Neuroefficiency aside, the Yuppie ideology of personal control is reinforced by biofeedback's promise of individual dominion over such bodily activities as hormone secretion, blood pressure, brainwave amplitude, blood vessel expansion, and the operations of the immune system. At their extreme Faustian limit, these enhanced biofeedback technologies promise complete voluntary control over the behavior of the body's every cell. This, clearly, is the alchemical dream of our high-tech times, bolstered by a promise of evolutionary self-determination that is no less Faustian; a good deal of New Age inspiration stems from the perception that the dawning of the Age of Aquarius brings humans the opportunity, for the first time in history, to take control over their evolution hitherto determined by genetic accident. In the new high-tech manifesto, the right to be intelligent is a primary human right. Science has made this possible; but modern science, in its dependence on large-scale technology, has developed certain "bad attitudes" that stand in the way of this goal. In this respect, New Agers want to avoid the Faustian mistake of making a pact with *externalized* technology, even if it means demonizing what might, under controlled circumstances, be liberating. The object, as Berman has suggested of the alchemists, is not necessarily to make gold, but to become golden.

Because of the community's suspicion about technologies that are external to the body's holistic orbit, the brain machines can thus be located on the outer margin of New Age legitimacy. Not only is the hardware more sophisticated than anything so far seen in New Age circles, but its dependence upon an external energy source (even when used simply to enhance biofeedback circuits) comes perilously close to violating the norms of "natural" bioenergetics, which only makes use of the inherent life energy of the human body. Here, we run up against some of the inviolate codes of New Age ethics, protective of the body–mind's own ability to control and accelerate its evolution without any external technological intervention. According to these codes, the natural laboratory of the body is almost self-sufficient and is capable of regulating its own experiments by the exercise of inner psychotechnologies, whose efficacy might then and only then be measured by external monitoring technology.

In the biocircuits that are more conventionally approved by holists, an external conductor may be used to complete the circuit and catalyze the flow of energy. The laying on of human hands is the most "natural" of these biocircuits—healers are taught to feel and interact with a patient's life energy. But there is also a whole range of nonhuman catalysts: crystals, magnets, and the use, in Eeman screens, of copper, silk, ceramic or micah.[12] More sophisticated technology takes us back to Edgar Cayce's wet-cell and impedance devices, Wilhelm Reich's orgone accumulators, and into the exotic fields of radionics and radioesthesia, all sensitive to the independent vibrational frequencies of the living organism. Each of these is an "appropriate" technology, because it merely acts as an accelerator of the basic processes by which the body heals itself.

The use of "appropriate technology" for the goal of personal transformation is one of New Age's governing principles; the concept was pioneered by the economist E.F. Schumacher to describe the advocacy of small-scale intermediate technologies for developing countries, whose cultures are threatened by the introduction of large-scale Western technologies. Although the term is not widely used in holistic circles to describe healing techniques, I find it quite relevant since it carries reminders about the lessons of postcolonial dependency that are usually glossed over in philosophies of New Age Orientalism, or in the new New Age patronage of traditional, archaic cultures.

In holistic terms, the inherent life-energy of the body is self-sufficient. The body, or the body–mind–spirit continuum, contains its own appropriate technology and does not require the intervention of external technological mediation to fulfill its own healing or evolutionary potential: left to its own devices, the body always knows what's best. At least one of the operative images is that of the body as a kind of electrical capacitor—an image continuous with nineteenth-century vitalism, as opposed to the more orthodox mechanistic notion that organic matter, far from harboring an *élan vital*, is no different in its capacity to conduct energy from nonorganic matter.[13] In general, however, the body's system of energy flow and harmony is not perceived simply as physical, but as the combined effect of the spiritual, social, and natural environments inhabited by the individual.[14] Alongside the archaic and the Oriental, there are also more postmodern images—the attempt, for example, to introduce elements of the cybernetic body familiar from modern molecular biology: the body must also be intelligent. Life energy is also, then, a conduit for carrying information; in the words of one advocate, the mind and body communicate "as a data cable carries transmitted information between two computers."[15] The New Age dream of establishing "communication" with our own cells swims into view. Contact with intracellular intelligence begins to take on the same recognizable features as contact with extraterrestrial intelligence. Talking with DNA is the corollary of talking with E.T. Inner space looks like outer space. If contact is possible, then contact *must* be made, the circuit of communication *must* be completed—if it can be done, it ought to be done. Why? Not only because people have a collective hunger for the sense of community, but also because in our information society the technical *fact* of communication itself is celebrated as an inherent good.

To follow this chain of interlocking imperatives is to recognize, in germ, the features of modern corporate communications ideology, which presents its will to wire up the world as if this were an evolutionary necessity. To "reach out and touch someone" is a progressive and necessary good, in and of itself. The service performed by the information giants in multiplying communication possibilities in every corner of the globe is usually posed as an unquestionable feature of the future. The question becomes: Who would not want it?, not, What do you want it for? Consequently, what is celebrated is the fact of communication; what goes begging is any public discussion of the resulting shape of the community that is wired up in this way.

This is probably not the place to develop any larger critique of communications ideology, although such a critique might explain how the technicist exterior of that ideology is supported by a strong organic core of therapeutic needs and communitarian desires that is equally relevant to the New Age cultures of contact. Consider, for example, the relatively independent New Age cults of UFO contacteeism and channeling. Each is a cult of intelligence in a knowledge society: the smartness of more highly evolved alien intelligences, on the one hand, and the ancient (*avant la lettre*) "wisdom" of channeled entities, on the other. Both cults, moreover, are utopian in ways that distinguish them from similar devotions within popular culture. In these cults we find little of the apocalyptic terrorism or theological dread that is such a winning affective feature of the commentary offered by pulp science fiction, horror, and fantasy genres on forces and intelligences beyond the realm of our sublunary knowledge. Indeed, one could argue that in these popular genres "higher intelligence" is justifiably perceived as a form of power and domination, and hence as something to be quite concerned about. By contrast, the New Age cults tend to view suprahuman intelligence as a familiar source of appropriate knowledge, far removed from the "alienation" effect invoked by scary monsters, ethereal brainwashers, and vengeful deities. To an outsider, the discourse of a channeled entity merely contains nuggets of unremarkable wisdom—which, however, has to be seen as both therapeutic and appropriate to the long-term self-organization of the everyday life of its audience. The "ordinariness" of this wisdom is a function of the trance medium's social constituency—a nonexpert, middlebrow, predominantly white audience, economically comfortable and sufficiently educated to respond to the bait of self-improvement if it is wrapped in good metaphysical taste. This audience is not positioned to respond well to the typically dystopian messages about knowledge-domination communicated by popular culture, primarily because of its class composition. For such an audience, the spectacle of higher knowledge is not likely to be associated with the terrorism of authority figures who have real power to command their lives in the workplace and in civil society; a more probable personification would be as a benign, respectful authority whose company they would like to cultivate, and whose actual presence they would like to entertain in their homes, or learn from in workshops. Equally distinct from the official knowledge technologies employed by intellectuals and experts, intelligence in the channelers' form of "perennial philosophy" (the metaphysical equivalent, if you like, of Masterpiece Theater) is a technology of therapeutic self-affirmation for people unsure about their place in the knowledge hierarchy, and who find a sense of communal purpose in responding to this teaching.

Discussion about the political impact of these "technologies" certainly ought to go beyond a mere sociology of knowledge castes. But I would be skeptical of any political criterion that champions one over the other; preferring, in other words, the popular culture over the middle-class cult. It has become acceptable to argue that seeds of "resistance" are more likely to be found in audiences faced with the affective edge of domination and alienation (however mediated) presented in certain forms of popular culture. From that point of view, the flat and bland sagacity of New Age consciousness is a political vacuum—no possibility of alienation, no possibility of resistance. This is not an adequate response, however, to what is essentially an amateur culture's attempt at self-determination, shot through with the contradictions raised by humanist pieties about self-completion and self-transcendence. Proper assessment of the politics of New Age consciousness, as I will argue later, requires a more thoroughgoing analysis of the problems raised by a purely humanist interpretation of the role that technologies of knowledge play in our culture.

For the time being, I want to conclude this section by reestablishing and reinforcing my points about the impress of modern communications ideology, specifically in the case of trance channeling culture. Although the range and context of channeled contacts is very broad indeed, in every case the medium basically channels communications from elsewhere, whether from deceased spirits, ancient entities, or the Universal Mind. This definition covers the whole specialized spectrum of skills ranging from spontaneous or clairaudient channeling, light or full trance, lucid dreaming, open channeling, automatism, sleep channeling, and clairvoyance, to the highly prized technical ability to read the Akashic records—a universal memory bank of knowledge that is accessed like a holographic video disk. A chronology of twentieth-century channelers would run from pioneers like Eileen Garret, Alice Bailey, the channelers of the *Urantia Book*, Edgar Cayce, and Jane Roberts's "Seth," to the more recent international flood of trance technicians: the Findhorn circle, the Columbia University psychologists who channeled *A Course in Miracles*, Elizabeth Clare Prophet (recent architect of the great Montana bomb shelter), J.Z. Knight's "Ramtha," Kevin Ryerson's multiple entities, Jach Pursel's "Lazaris," Jessica Lansing's "Michael," Penny Torres's "Mafu," Thomas Jacobson's "Dr Peebles," Ruth Norman's "Tesla," Shawn Randall's "Torah," Ken Carey's Starseed Transmissions, Sanaya Roman's "Orin," Jamie Sams's "Leah," Daryl Anka's "Bashar," Alan Vaughan's "Li-Sung," Maurice Cooke's "Hilarion," Benjamin Creme's "Maitreya," Carolyn Del La Hay's "Ramala," Gabriel Wittek's "Jesus," and Alice Anne Parker's "Menos."

Skeptical debunkers of this culture tend to fixate on the (un)authenticity of the source, while religiously minded critics denigrate the mixed humanist/transcendentalist content of much of the channeler's discourse. More significant to me, however, is the culture's celebration of its ability to resolve the technical problems of communication. In this respect, channeling culture is continuous with the tradition in the history of religions of obsessive attention to the technological trappings of earthly mediation. If high technology today has come to take the place of fetishes, symbols, icons, wafers, and scriptures, the technology still has to be culturally appropriate for communication to be successfully established. In the case of channeling, the problem often lies in finding an efficient interface for the translation of "energy"—the constituent personality essence of channeled entities—into physically apprehensible communication. Here, for example, is how "Lazaris" describes how communication, originally in the quasi-digital form of "blips and bleeps," takes place between his personality and his trance channel, Jach Pursel:

> Downstep, downstep—it's a downstep generator in electrical terminology—to downstep this particular vibration to where it reaches a certain density. Then those particular bleeps and blips, we can take that mass of energy and align it because it's dense enough to be able to be aligned. . . . And it steps down to the Astral Plane into the physical. And then the antenna which is the channel—for that's what the channel is, an antenna—it comes into him, the blips and bleeps. They are amplified and they come out of here [points to Pursel's throat]. You hear the blips and bleeps and they vibrate against the ears and send a pattern of signals to the brain, and the brain is where these vibrations become words. . . . We would suggest here that it is because of the particular alignment of the energy that it comes through as a particular accent. . . . There is an interface with light. What happens is that the vibration is downstepped into the various planes. It comes to a certain place where it interfaces. That point might be called a conversion from love to light. Prior to the point that it is light, it is love. At a certain point that love translates into a vibrational frequency that is particle and wave simultaneously. For us to interact with you there must be a *you*. The love that we are gets downstepped to the point where it can be systematized and in sequential order, and thus is the love aligned. And then it is downstepped further until it meets that level where it converts to light. It converts to wavelengths, wavelengths of light, where it becomes measured. It converts to particles when the frequency hits this instrument, which is what the channel is—vocal chords are an instrumentation. When the waves of light hit the instrumentation, they are converted simultaneously, instantly, into particles—particles, frequencies of vibration, which then are amplified.
>
> And that's the quantum. That's the mysterious leap that occurs. And by the way, scientists have now seen the release of the electron without motivation. Without cause, they've seen effect . . . they've got conclusive evidence of the quantum.[16]

Such detailed attention to the technical difficulties of establishing contact underscores the fact that what is being valorized in this account are the technologies of communication in and for themselves, raised here almost to the level of metacommentary. And where is the technological sublime in such a discourse? Godhead residing in the interfaces between love and light and sound.

In the course of his explanation, "Lazaris," like most other channeled entities, coyly intimates that his knowledge is "beyond" the science of the day: it not only incorporates the folk wisdom of preliterate cultures, but also functions as an avant-garde source of information about future scientific research and discoveries. Because of its commitment to alternative access to knowledge, the channeling community is unable to draw directly upon the accredited word of scientific authority; preferring, as Lazaris does here, to invoke that authority as matter-of-fact corroboration of the entities' wisdom. Authority on scientific matters is sought elsewhere: either in the wisdom of all ages, which knows the future but cannot tell it, or in the benevolence of alien intelligences, whose willingness to impart information about advanced technologies is either thwarted or unfailingly misunderstood in their earthly "visits." Perhaps these difficulties and breakdowns in communication, these obstacles in the path of technology transfer, are a necessary cultural element of narratives of "contact" that might otherwise bring to mind the long, barbaric history of colonialist associations with the idea of making "contact." After all, smooth technology transfer from aliens would much too closely resemble a colonial narrative, with white Westerners now in the position of the colonized receivers. To prevent the narrative from taking on a colonial aspect, channeling presents a scenario in which *existing* technology must be arranged, developed, and utilized in the name of higher communication, never too directly personified, and always as a way of establishing on-the-ground technical control over the channel of communication.

Among the more apposite examples I have come across are the transcriptions by a group at Rock Creek Research and Development Laboratories in Louisville, Kentucky, entitled *The Ra Material: An Ancient Astronaut Speaks* and published in 1981 after two decades of experiments in establishing contact. Here, the traditions of ufology and trance channeling seamlessly converge. Ra, members of an ancient alien race, came to Earth 11,000 years ago in order to accelerate, through technology transfer, the evolution of the human species. They visited the Egyptians, and the Incas, and helped to build the pyramids. Too weird to be accepted as gods, and too highly evolved to communicate efficiently, they retreated to an extraterrestrial monitoring position from whence their communications can be picked up by those "alienated" earthlings whose vibratory domains are sensitive enough to be aligned in harmony with that of Ra. As a "sixth-density social complex" that is three evolutionary cycles ahead of us (i.e. millions of years), it is no surprise that Ra still has immense difficulty in establishing communication with us. The process has to be helped along by laborious and inventive fine-tuning of the receiver technology on the part of the human channeling group.

For contact to take place, a "human" instrument has to be aligned with the head pointing 20 degrees north-by-northeast, and then "precisely tuned" to the narrow band vibration that Ra has selected for their communicating frequency. All subsequent communications are distorted, but some are intelligible if the receiving instrument is properly "tuned." Success depends entirely upon ceaseless attention to details of arranging and repositioning the elements of the receiver technology. Most of the materials involved in preparing the human antenna are traditional, low-tech Christian trappings: a white candle, a white robe, "a small amount of cense, or incense, in a virgin censer," "a virgin chalice of water," and a copy of the Bible, the book "most closely aligned with the instrument's mental distortions."[17] Above all, however, it is the human body itself that acts as the receiver. In the more hard science-oriented fringes of the broad UFO subculture, preparations for signal reception involve less appropriate technology—advanced satellite hardware that is often so large-scale as to dwarf, if not entirely exclude, the role played by the human body in completing the contact circuit of communication.[18]

ZEN AND THE ART OF MAINTAINING POWER

In my discussion of the brain machines, I argued that their reliance upon an external energy source jeopardizes their holistic status as natural biocircuits. What redeems them, however, is their claim to dialectically reorganize the brain, through synchronizing left-side and right-side, at a higher level of complexity. Behind this Hegelian move lies a host of New Age claims about neuro-evolution, quantum leaps, and negentropic behavior premised on the lofty speculations of the new physics. On the horizon lies a rapprochement, we are often told, between a metaphysics that is no longer anti-science and a scientific rationality that is no longer impersonal and alienating.

In looking at biocircuits and channeling, respectively, I have visited two ends of the New Age spectrum of images of the body. In the former case, the mind–body is a fully enclosed, self-communicating system of energy and information; in the latter case, the mind–body is a conducting medium, a receiving antenna or a transforming station in a much larger circuit of energy and information. A philosophy that wants to contain and make sense of both models is obliged to resolve one of the oldest problems in metaphysics—addressed, for example, by Leibniz's "monadology"—of communication between a closed, holistic system of consciousness ("windowless" in Leibniz's terms), on the one hand, and the universe of intersubjective consciousness on the other. New Age Orientalism offers many traditional solutions to this problem, mostly in the form of cosmic energies that are commonly circulated through the body. It is from modern brain science, however, that New Agers have drawn the most competent explanatory models for a new cosmology with science as its sustaining cultural core.

The most serviceable model comes in the form of the holographic paradigm, drawing upon the claims of Karl Pribram, David Bohm, Kenneth Pelletier, John Battista, and others, that the brain's deep structure is holographic: the information from which consciousness works is stored all over the brain in holographic fashion and is accessed accordingly. Once the brain's ecology is understood as holographic, the principles of isomorphism and synchronicity, from brain to brain, come into play. Sensory reality appears as a relatively stable representation, but is projected holographically from a point that is, in principle, beyond time and space. If the universe itself becomes a master hologram, all of reality can then be recovered from its smallest portion; each brain incorporates the universe's information. Holism is thereby established at all the implicate levels of experience.[19]

A number of philosophical problems are addressed by this paradigm. It not only establishes a permanent, fluid ground for intersubjective communication, but also allows for a more socially equitable overall distribution of energy than the *karmic* universe of retributions and rewards. Just as a formalist might argue that the politics of atom-smashing somehow equates to an attack on the centered Cartesian subject, so holism's proponents see the unified holographic field of perceiver and perceived as a leveling critique of the privileges of subjectivism. Such a field accommodates "mystic experience" not as a contingent or aberrant encounter but as a rational apprehension of the conscious holo-movement of sensory reality.[20] More pragmatically, the holographic paradigm provides a language for, and thus a new way of representing, the uncertain and unstable world of fields and particles that has been the focus of the new physics for some decades now.

If metaphysicians no longer habitually find themselves placed in the anti-science corner, it is because theoretical science, in the wake of quantum physics, has shattered the intellectual security of the mechanical picture of discontinuous time, space, matter and objectivity. Quantitative rationality—the normative description of scientific materialism—can no longer account for the behavior of matter at the level of quantum reality. This is particularly true of the particle–wave duality of subatomic physics, in which particles that do not behave like particles exhibit a wave or field-like movement. In addition, the Heisenberg principle has established that the measuring observer inevitably becomes part of the experiment itself; objectivity and subjectivity are then emptied-out categories since there is no quarantine space for testers or their measuring instruments, whether material or mental.

Consequently, theoretical physics has moved on to ground that has been traditionally claimed by metaphysics—organicist rather than mechanistic, and transcendentalist, even spiritualist, rather than materialist. Newton, Galileo, Locke, Bruno and Descartes are displaced by Lucretius, Leibniz and Whitehead. Physics has become a field of knowledge about the unknowable, taking on many of the epistemological aspects of a high speculative religion, while the new "impossibility" of its object of knowledge aligns it with fields, like psychoanalysis and parapsychology, normally excluded from the purview of the scientifically constituted. For many

New Agers, there is good reason to conclude that this marks the end of the long conflict between scientific rationality and the realm of cosmology and spiritualism inhabited by metaphysicians. Weber's "disenchantment of the world," fomented by rationalism's separation of man from nature, is held to be over, and a wealth of observations about mysticism and religion from famous physicists—Heisenberg, Einstein, Bohr, Schroedinger, De Broglie, Jeans, Planck, Pauli, Eddington—are customarily cited in order to illustrate the new fraternity between physics and mysticism.[21] From the metaphysician's perspective, the paradigm-shaking effect of developments in quantum physics can and has been read as a sign that science is finally coming to its (extra-ordinary) senses, and is now entertaining less material visitors who were once personae non gratae in its house—visitors whose exclusion was once the mark of science's constitutive identity.

Scientists starting to talk religion may well be one bizarre symptom of the crisis of rationality, and a harbinger of the day when modern science can make sense alongside the oldest recorded thought of the world's religions. More to the point, however, is the spectacle of metaphysicians not only speaking the language of science, in full deference to its own inbuilt deference to accredited authority, but also competing for the power and prestige associated with that language in our knowledge hierarchy. Metaphysicians have seldom found themselves in a position to enjoy the legitimacy that derives from talking science. By that same token, the spectacle of scientists speaking about Godhead in the Stanford linear accelerator is not so much a sign of rationality's grip upon the object world relaxing as it is the sign of a new level of power—the technological sublime—achieved for a rationality that can now conceal, better than ever, the social and material conditions of its existence from the layperson. The rhetoric of the sublime further excludes those who are interested in science's democratic accountability.

At least two New Age books by scientists—Fritjof Capra's *The Tao of Physics*, and Gary Zukav's *The Dancing Wu Li Masters*—became bestsellers in the seventies by pursuing the thesis that the new world-view of quantum physics was in harmony with ancient Eastern metaphysics. As theoretical physics grappled with the problem of representing its quirky, or quarky, new knowledge in the linear form and languages of Western rationality, Capra and Zukav argued that the paradoxes it encountered resembled those chosen by Zen masters and the like to represent the kinds of mystical experience recounted in the philosophies of Hinduism, Taoism, and Buddhism. Stumped by the problem of describing wave–particle duality or quantum fields, the high-energy physicist was encouraged to take comfort in the knowledge that Zen philosophy communicates its teachings without rational explanations, instructions, or axiomatic knowledge. Capra expressly points to the long years of training shared by the apprentice physicist and the Zen student in pursuit of their highly specialized crafts. Both are initiated into serious enquiry about the nature of the universe, and both become the curators of advanced knowledge about secrets that are inaccessible to the uninitiated.[22]

For those who want scientists to make their work more accountable to the nonexpert, Capra's analogy is, in every respect, a step in the wrong direction. Far from demystifying the work of science, it elevates the scientific vocation beyond the status it already enjoys as a secularized Western priesthood. Ordinary language and everyday rationality are revealed as inadequate, archaic, and therefore redundant media of communication. When the words of the physicist begin to sound like a koan, the aim of explicating science in the vernacular to a nonexpert audience has been abandoned. What emerges here is not a picture of benevolence and humility—Western science kowtowing to the humanism of non-Western wisdom—but the contrary: a sign of the further empowerment of the scientist, allotted the task of pursuing reason's higher mysteries, released from any obligation to speak of them in an intelligible language, and thus sent out in the emperor's exotic new clothes to proclaim truths that few people will be capable of understanding.

New Age affirmations of the vanguard role of skilled metaphysical consciousness do not, of course, always depend upon the support of Oriental thought. In New Age literature, metaphors that draw upon Western technologies of transformation are common enough, especially when those technologies can be internalized as figures of flight:

> A mind not aware of itself—*ordinary* consciousness—is like a passenger strapped into an airplane seat, wearing blinders, ignorant of the nature of transportation, the dimension of the craft, its range, the flight plan, and the proximity of other passengers. The mind aware of itself is a pilot. True, it is

sensitive to flight rules, affected by weather, and dependent on navigation aids, but still vastly freer than the "passenger" mind.[23]

> You're sitting in the left seat of 174 tons of machine, waiting for takeoff clearance. You switch to departure frequency. A staccato voice in the control tower says, "Flight 972, you're cleared for takeoff. Blue Angel 7, continue to hold for clearance." You glance over to your co-pilot. "Let's roll." The whine of the engines begins its crescendo in your headset. The power of the bird is unleashed as you push the throttles forward with a sensitive movement gained from thousands of hours in the seat. . . . Your nosewheel begins to lift off the ground, pointing upward into the sky. In a few seconds, the rest of you will follow. You're airborne . . . rising into the new understanding, traveling into Superconsciousness, into *Flightlevel: Freedom*.[24]

Much more rare, however, are attempts to focus honestly on the specifically Western contradictions that arise from the relationship between dynamic consciousness and external technology. Almost unique in this respect, at least in my experience, is Robert Pirsig's novel *Zen and the Art of Motorcycle Maintenance*, which was probably one of the last "sixties" cult books to enjoy high legitimacy on the syllabi of hip college professors. Published, like *The Tao of Physics*, in 1974, Pirsig's book, unlike Capra's and in spite of its own Orient-directed title, does not lead the reader away from the gridlock-plagued thoroughfares of Western rationality into the promised land of the lotus. References to Eastern enlightenment put in an occasional token appearance with whimsical rather than mystificatory effects:

> The Buddha, the Godhead resides quite as comfortably in the circuits of a digital computer or the gears of a cycle transmission as he does at the top of a mountain or in the petals of a flower. To think otherwise is to demean the Buddha—which is to demean oneself.[25]

For the most part, the novel confines itself to the contradictions of its own culture, divided in the aftermath of the sixties by widespread challenges to its prevailing rationalist tradition. It tells the story of a brilliant young English instructor whose pursuit of the "ghost of rationality" throughout Western philosophy leads him to a traumatic break not only with the education system but also with his own psychic stability as defined by the other, medical institutions of rationality. His attempts at self-healing involve a process of reskilling in rationality—learning the *art* of technology through motorcycle maintenance, an art that had been forgotten as external technology increasingly made his generation "strangers in [their] own land."

Zen and the Art of Motorcycle Maintenance is tragically set against the backdrop of sixties technophobia (the countercultural revolt against reason, order and materialism), represented in the characters of the narrator's traveling companions, whose hatred of technology obliges them to know little and learn even less about the art of motorcycle maintenance, and whose attitudes the narrator finds to be finally "self-defeating." By contrast, his own tender devotions to the machine and its component parts convey a spiritual relation to technology habitually denied to people living under the influence of instrumental reason. "A motorcycle," he muses, "is a system of concepts worked out in steel," governed by the classical "mastery of underlying form." In this respect, his motorcyle is a figure for all Western technology. To resist the labor of understanding the rationality that is technology is to cede any individual control over the power of technological rationality. There is no specialized knack involved, since machines entirely obey the laws of reason. Mastering these laws amounts to a triumph for technological humanism.

Pirsig's novel was an admirable response to the demonizing of technology in the sixties as a "mindless juggernaut," a mechanical monster which (in bureaucratic complicity with what Charles Reich, in the canonical *Greening of America*, called the "machine rationality of the Corporate State") was seen to be wreaking uncontrolled havoc on our natural and mental environments alike. Even if its model of liberation was solipsistic (the lone, tortured consciousness of his narrator, failing, finally, to establish a relationship with his son) and prefeminist (motorcycle maintenance was too dirty for women to be much interested in it), Pirsig's book presented an alternative to the romantic disavowal of modern technology that was all too common in countercultural literature and thought.

The larger tragedy of the novel, however, lies in the story it tells about the narrator's failure to sustain his critique of rationality within the institutional context of the education system. In hunting down the ghost of rationality, he destroys too many of his own intellectual resources—irredeemably shaped by rationality—to communicate and contest within an institution founded upon that very concept. Although it rejects any esoteric flight into the many arms of the Buddha, Pirsig's novel finally reverts to the romance of the "open road"—the traditional site, in North American letters, for self-reliant masculine souls in flight from the institutions of reason (the university), correction (the prison, or psychiatric clinic), mysticism (the monastery) and domesticity (the family). In confining its lessons about technology to the realm of everyday contact and understanding, the novel advocates an "appropriate" technocultural milieu. On the other hand, its depiction of an institutional failure to contest existing forms of Western rationality may have been an honest response to the contradictions of its day, but is of little help, finally, in counteracting the contemporary move of Capra et al. to extend to scientists and technologists the privileges of non-rational insight. Dropping out of an institutional system is no match for the ultimate promotion—a lab in the sky.

NON-TYRANNICAL NATURE

In its search for ever larger structures of wholeness and unity beyond the individual body–mind, New Age metaphysics also displays the authoritative influence of modern biologists' rethinking of entropy and evolution, and thus of nature itself. Nineteenth-century thermodynamics had shaken the classical legacy of mechanism's ordered, inert universe by suggesting that energy was leaking out and that the universe's clock was running down. In contrast to this picture of an entropic system going from order to disorder, Darwinism showed that biological and social systems were negentropic systems; progressing, if you like, from simple to complex order. Modern biological systems theory has suggested that order can now come from chaos, to cite Ilya Prigogine's famous formulation. Despite the irreversibility of nature's processes, open systems that exchange energy with the environment can be seen to maintain an order and stability, even in far from equilibrium states. Self-organizing, dissipative structures, constantly being transformed through feedback loops with environmental energy, do not obey the classical laws of entropic behavior for closed systems. These structures increase in complexity with the input of feedback, constantly interacting with the energy exchanged. The result? Processes, practices, and formations, rather than the substantive organisms and closed circuits envisaged by classical mechanism. Disorder, disequilibrium, and diversity coexist in a natural order that is multiple and irreversibly temporal.[26]

This is a quite different conception of nature—nonlinear, unstable, and stochastic—from that established by Locke, Newton and Descartes. It no longer resembles the dependable referee on and against which one can verifiably test empirical propositions. Nor is it a picture of nature as a tyrannical determinant of every final cause of human and social life. Above all, it offers no secure vantage-point for nature's observers, no guarantee of immunity for its ethnographers. New Age subscribers to the holistic promise of a higher organic order have seized upon this renewal of faith in non-entropic evolution. The New Age, as it is conventionally presented, stands breathlessly on the brink of a collective leap in evolution—a major upgrade for the species as a whole. How to accelerate this evolutionary leap without violating the "natural" way of things (it is no longer apposite to speak of the "laws of nature")? This is the question directly posed to technology itself, the noisy, polluting vehicle of science's ideas of progress. How "external" do technologies have to be to fall outside the orbit of the "natural"? And how is consensus reached on drawing the line of demarcation?

In addressing these questions, New Age practitioners have placed their evolutionary faith in what are commonly known as the "psychotechnologies." These extend from the biofeedback machines at the hardware margin of the spectrum, to shamanic and magical techniques at the visionary end. It is a vast and accommodating spectrum, however, stretching across the whole range of holistic therapies, bodywork disciplines, paranormal activities, and alternative religious practices. Even a selective list of the numerous holistic psychotechnologies would include reflexology, rebirthing, creative visualization, aromatherapy, flotation, acupressure, actualism, naturopathy, Bach Flower therapy, holotropic breath therapy, numerology, herbology, connective tissue therapy and deep tissue bodywork, personology, iridology, muscle therapy, Touch for Health, polar energetics, Neo-Reichian release, orthobionomy, biomagnetics, flotation, Reiki, mentastics, chakra therapy, past life regression therapy, magnetotherapy, kinesiology, Rolfing, colon healing, stress management, ecstasy breathing, craniosacral therapy, body tuning and vibrational medicine, quantum healing, polarity therapy, chelation therapy, phototherapy, neurolinguistic programming, the Trager approach, Aston Patterning, the Alexander technique, the Radiance technique, the Rubenfield Synergy method, hakomi, ohashiatsu, Jin-Shin Do.

At the common core of the psychotechnologies lie the principles of self-help, self-healing, self-equilibrium, and self-transcendence. Many of the holistic practices are associated with premodern or non-Western cosmologies for which healing technologies are not materialist processes. In his 1975 essay, "Art as Internal Technology," José Argüelles, who would later become the architect of Harmonic Convergence, put the case for a redefinition of "technology": "the archaic is a counterpoint to technology which is an instrument of externality . . . archaic techniques constitute an internal technology."[27]

The broad community of holistic practitioners ranges from private healers operating within confidential networks to fully licensed, professionalized practitioners operating within the restraints of community or state standards.[28] While training is important, the underlying principle is an amateurist one—everyone has the potential to become the engineer/architect/designer of his or her own environment. Reskilling oneself in the arts of the psychotechnologies can be seen as a way of reappropriating folk skills from the experts that were once everyday knowledge. The recent revival of traditional medicines in countries like India and China, where rational, cosmopolitan medicine was already well established, has demonstrated that nine-tenths of what Westerners think of as "medicine" is within the non-professional's grasp. The flourishing interest in non-cosmopolitan medicine is also inspired by social critiques of the medicalization of health.[29] Biomedicine, "of all industries, the most wasteful, polluting, and pathogenic,"[30] has been held responsible for an explosion in iatrogenic illness—the diseases generated by institutional medicine (poisoning, infection, and unnecessary surgery in hospitals, which are the site of most accidents)—and the general appropriation of people's bodies by professionals dependent on megatechnology.

In recent years, the legitimacy of medical professionals has been further eroded, as popular consciousness absorbs the more general social critique of biomedicine's "inhumanity" and lack of caring for the patient. In a profession where the orthodox faith in the (large) technology fix has often led to consequences "worse than the disease," the multitude of alternative health therapies (the women's alternative healthcare movement, for example) has made its biggest inroads. The professionals' bad attitude is often caricatured in the following way: biomedicine sees and treats the body as a functional machine that occasionally breaks down, and for whose dysfunctions a physical cause and remedy can always be found by the repairman, the doctor. In addition, biomedicine is seen to have "failed" in its treatment of chronic, degenerative diseases like cancer and cardiovascular illnesses, thought to be primarily caused by social and environmental conditions; its miracle drugs often generate further problems in their side-effects and consequences. The scandal of the AIDS crisis and the particular social forms and pathways taken by this virus have exposed the extent to which medicine is a fully political process, saturated by homophobic, sexist, and racist prejudice. In general, biomedicine's role in contributing to health is actually rather small when compared to social and environmental conditions, and it is precisely because it has neglected these factors that its closed, mechanistic model of the body as a system of efficient causality has been seen as insuffIcent.

Medical professionals have absorbed these criticisms to some extent. Here is some anecdotal evidence from my own experience. Not long after moving to the United States from Britain, I visited a dentist to inquire about my first symptoms of gum decay. In an attempt to allay my anxiety, the dentist told me that it wasn't my gums that I should be worried about. "Look at it this way," she said. "It's not your gums that are decaying, it's your body that's decaying." She was, I suppose, appealing to the knowledge I was supposed to have about the irreversibly entropic nature of my body, here showing the "natural" signs of advancing age. But I was unaccustomed to meeting dentists who did not simply fetishize the contents of my mouth, and her advice, even if offered tongue-in-cheek, did give me cause for alarm. Having recently moved from a society where responsibility for the upkeep of my teeth had always been seen as the obligation of the welfare state, this dentist's implicit suggestion that bodily maintenance was now my own economic—even moral—responsibility took on ominous overtones for me. Although my dentist was not a holistic practitioner, she had made reference to a holistic idea about the interdependence of bodily functions. In retrospect, I realize that a more serious holist would not have placed any such emphasis on the entropic process; I have discovered that a wide range of alternative psychotherapies promote techniques, usually involving hypnosis, that claim to redirect the body's energies to heal tooth and gum decay and thereby partially reverse the entropic process.

Many holistic health practices, in fact, are unequivocally in the business of *reversing* entropy. In contrast to the paternalistic diktat of the health professional, holistic therapies have set up shop on the basis of faith in the body's self-healing faculties. There, the health professional's obscurantism is often replaced by an apparent voluntarism: "Health and disease don't just happen to us. They are active processes issuing from inner harmony or disharmony, profoundly affected by our states of consciousness. . . . This recognition carries with it implicit responsibility and opportunity. If we are participating, however unconsciously, in the process of disease, we can choose health instead."[31] The AIDS crisis, for example, far from being seen as the result of criminal mismanagement on the part of politicians, bureaucrats, drug companies, and physicians, is viewed in certain New Age circles as the result of negative thinking: it is not the virus that makes us ill, it is our negative overreaction to the virus. It is no small irony to see, in the thick of AIDS activists' demonstrations and protests, New Age banners proclaiming "The Healing is Happening," commingling with more militant slogans that castigate the "genocidal" effects of government health policies while insisting that "Health Care is a Right." There is no doubt that some PWAs have turned to holistic remedies more out of desperation than because of any philosophical opposition to drugs as such. Nonetheless, the holistic spirit of looking after oneself in one's own environment seems to be at odds (though not irrevocably so) with activists' demands that medicine be redefined as a social process in which the people most affected should have the right to participate at all levels: decisions about government policy and funding, pharmaceutical development and testing, and inexpensive access to care and treatment.

In fact the immune system, cast into the spotlight as a leading health actor by the AIDS crisis, has become one of the favored metaphors for the New Age body's capacity to take care of itself. If orthodox biomedicine treats the body as if the mind were not there, then the body–mind–spirit continuum that is central to the New Age concept of health appeals directly to transcendentalist codes of self-reliance. The physical is the least important part of holistic treatment, since it is seen primarily as a symptom of disturbances in the individual's spiritual and social environments. The patient is asked to make a "commitment" to his or her health, and the cure involves a good deal of self-analysis, for which the healer acts as a catalyst in restoring the natural balance of the body–mind–spirit's energy economy.

More often than not, however, the argument for holism is reduced to the idea that health must be actively pursued by a person and not simply underwritten by a bureaucratic commonweal. What is left begging is any question of a socialized healthcare system that can properly accommodate holistic healing practices. It takes little political ingenuity to see that the former perspective transforms the business of health into an entirely privatized affair, decoupled from the welfare state and hostile to any notion of intervention on the part of social agencies. This is partly a consequence of the fierce independence of the holistic community, and partly the result of the *sub rosa* legal conditions under which many of its practitioners are obliged to operate. For holistic healers, concepts like the "individual" and the "body," along with the cultural beliefs that support them, *mean* something quite different from equivalent concepts as defined in orthodox biomedicine and in the atomistic individualism of liberal ideology at large. But these practices, and the particular shape of the philosophy of health to which they subscribe, have not evolved in a social vacuum, immune to the general ideology.

In many respects, the holistic picture of the body–mind–spirit as an efficiency state of equilibrium tends to reflect or express the ideology of the "natural" self-regulating organism of the free market economy that has again, during the last two decades, come to prevail over the troubled Keynesian body of liberal social democracy. Logical flaws in the bureaucratic technology of "external" statist solutions have brought the welfare house down; now, all responsibility falls upon the more "natural" economy of individualism. To drive this point home, I suppose we could say that what we saw under Reaganism and Thatcherism was the revival of a "homeopathic" economics, where the principle of supply-side investment acts as a kind of biofeedback input, a trickle-down stimulant for accelerating the dynamic health of the free market system as a whole.

The new "insurgent" therapies reflect this recent ideological reformation just as surely as they trade on the popular suspicion and resentment of the institutions of allopathic medicine. Only a diehard rationalist would be surprised to find that an alternative health paradigm appealing to personal values and self-reliance has won over significant portions of popular consciousness in recent years, partially redefining the given wisdom, or common sense, about our bodily and spiritual relation to the natural world. But the insurgents are often just as blind to the incipient conservative meanings attached, in the culture at large, to the philosophy of individual responsibility that they have espoused. In the eighties, conservative forms of populism were quite successful in collapsing the perception that individuals *count* into the laissez-faire principle of self-reliance as the foundation of the social good.

In her book *The Whole Truth*, a critique of the alternative health movement, Rosalind Coward sums up the reductive effects of the new social philosophies centered upon the individual body:

> This new concern with the body is a place where people can express dissatisfaction with contemporary society and feel they are doing something personally to resist the encroachments of that society. Indeed, so strong is the sense of social criticism in this health movement that many adherents proclaim that they are the avant-garde of a quiet social revolution. Yet the journey to this social revolution is rarely a journey towards social rebellion but more often an inner journey, a journey of personal transformation. The quest for natural health has come to be the focus of a new morality where the individual is encouraged to exercise personal control over disease. And the principal route to this control is a "changed consciousness" and changed life style. With this emphasis on changing consciousness have come all the fantasies and projections associated with religious morality, fantasies of wholeness, of integration and of the individual as origin of everything good or bad in their life. And with these fantasies there has mushroomed the industry of "humanistic psychotherapies" emphasizing the role of the individual will-power in making changes.[32]

Coward goes on to argue that health has displaced sexuality as the new privileged discourse of bodily truth and inner essence. Health has become the new moralizing category for power to exercise its hegemonic oppositions. Bad health is a sign of the moral failing that only recently was attributed to repressed sexuality. As Coward points out, this new good health doctrine recycles many of the Christian dualist elements of a doctrine of salvation, in which the thirst for self-purification reappears unassuaged. The perpetual, paranoid maintenance of a cleansed, purified body, immune to all sorts of external pollution, tends to feed into a social philosophy saturated with the historical barbarism of the politics of quarantine, natural selection, and social apartheid.

A similar argument about the displacement of "sexuality" by "health" could be addressed to the metaphysical holism of the body as conceived by the alternative therapies. The holistic body bears little resemblance to the fragmented body favored by, say, psychoanalytically oriented poststructuralist thought, although both philosophies are equally critical of the ego-centered Cartesian humanist tradition. In contrast to the fractured life of the poststructuralist body, shot through with the disordering effects of the unconscious, the healthy New Age body is an efficient, smoothly circulating system of energies, resonating, on a higher plane, with the natural frequencies of the global/astral body itself. The post-Freudian explanation of bodily "nature" is one that includes the troublesome, contingent narrative of a psychosexual history, inseparable from the narratives of socialization. By contrast, the New Age body, inasmuch as it is aligned with "nature's ways," is quite ahistorical; it is therefore immune, and often strictly opposed, to any conception of "nature" as a social construct. Appeals to history, when they appear, are usually timeless appeals to cosmic or mythological history.

In this respect, New Age advocates of the "natural" have developed some affinity with certain radical feminist and eco-feminist ideas about woman-as-nature. As the traditional keepers of intimate knowledge about health and the body, women are preeminent among New Age adherents. The most popular New Age exponent of this feminist philosophy is Lynn Andrews, the bestselling author of spiritual adventure fictions like *Medicine Woman*, *Windhorse Woman*, and *Star Woman*, whose shamanic initiation seminars ("Into the Crystal Dreamtime") attract thousands around the country:

> I'm spending my life writing about experiences with ancient women that are part of the Sisterhood. My work has to do with becoming a bridge between the primal mind and white consciousness; we as people—the White, the Black, the Indian, the Asian, all of us have become part of the mechanized, materialistic society—have lost our sense of Mother Earth and Nature. We have lost that wilderness of our soul. My work has to do with bringing us back to that mother, which is Mother Earth, the feminine consciousness that is so lacking on this planet today.

Andrews's books are a women's version of the "magical autobiography" popularized by Carlos Castaneda, and are the romantic, neocolonialist vehicle for exotic Harlequin-like adventures in exotic places, usually aided and abetted by the appropriate technology of psychic Arabian thoroughbreds. A more grounded version of this women's genre appears in the bestselling autobiographies of Shirley MacLaine, whose name is virtually synonymous in the popular mind with the public face of New Age consciousness.

NON-SOCIAL NATURE

The view that "nature" is a final arbiter of human behavior is quite central to the whole range of psychotechnologies. To take an example from the non-metaphysical health fringe, I will cite a few of the arguments presented in a critique of the "modern toilet," which accompanies promotional literature for "The Welles Step," manufactured by Welles Enterprises of San Diego. The Welles Step, a plainly fashioned pedestal or stool that allows people to squat while sitting on the lavatory, is promoted by appealing to nature: "People were intended to squat. They squatted throughout history. With this posture the abdominal wall and bowel are supported as we bear down. This is nature's way." The manufacturer promises to remedy "incomplete elimination," kinked bowels and fecal stagnation, the entry of toxins into the bloodstream, hemorrhoids, varicose veins, and diseased colons and bodies. All of these claims answer directly to the holistic model of a smoothly circulating bodily system, where the cleanliness of the colon, in particular, is a major reference point.

In a technical paper by Dr Welles, entitled "The Hidden Crime of the

Porcelain Throne," the author explicates his medical discovery of the "ileocecal valve," which regulates the flow of digestive contents from the small into the large intestine. Midway through the paper—a narrative report on his research findings—Dr Welles announces: "I then came to suspect the Western toilet as a causative factor in the all too commonly dysfunctional state of the ileocecal valve found in my patients." The appearance of the epithet "Western" is like the mark of Satan for the holistic health community, which systematically conceives of us as victims rather than beneficiaries of specifically Western technology. What follows almost inevitably in Dr Welles's paper is an appeal to the wisdom of the older, squatting (non-Western) civilizations, punctuated by further elaboration of his research findings. I am no authority on the digestive system, and so I cannot report on the validity of Dr Welles's medical theories. I can, however, point to what is clearly missing from his analysis—any description of the historical or ideological conditions under which immaculately white porcelain toilet technology was developed to demarcate squatting from non-squatting populations, and thereby create, if you will, an international division of excremental labor.[33] The vestiges of this division are still clearly evident in European cultures stratified along North/South and internal class lines, and in postcolonial cultures where imported Western toilet technology still exercises its inscrutable power over squatters. Welles's only gesture in this direction is his observation that although the modern toilet was "well suited to the sedentary Westerner," it was created "with absolute disregard for the anatomy of the human body." On the latter point of anatomy, I am, again, not qualified to say whether he is correct, but the former point leaves begging the question of Western "upright" sedentariness, with all of its connotations of moral and racial superiority. Notwithstanding Dr Welles's moral resentment, technologies are seldom developed simply with human, anatomical functionalism in mind. Their architecture embodies existing values—heavily socialized, heavily politicized—in a particular culture at a particular historical moment. The highly technological concept of a "correct" or "natural" human posture is itself a culturally relative idea for which no universal norm can be assumed.

There is, of course, no reasonable criterion by which we could expect a medical entrepreneur like Dr Welles to include the social history of the

The Welles Step

PATENT PENDING

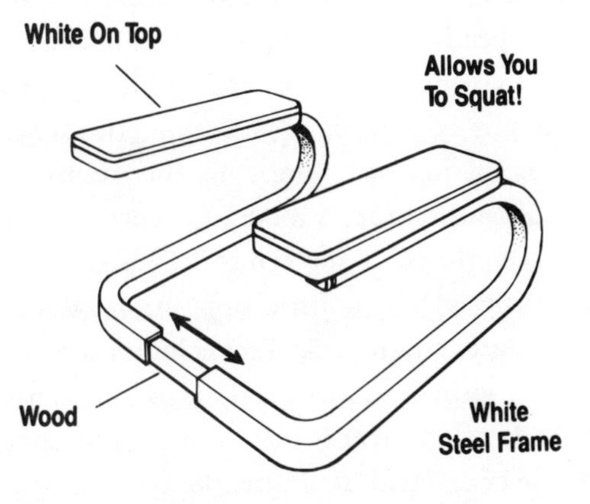

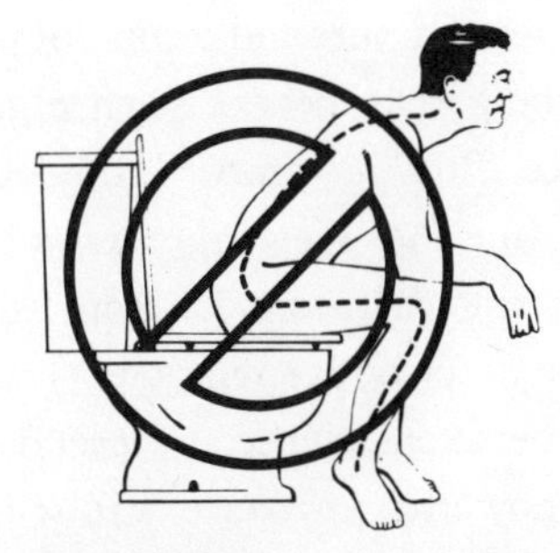

Stores Out Of The Way

The Modern Toilet was a great mistake. It leaves these two areas of the abdominal wall and bowel unsupported as we bear down. (See red area on illustration below).

People were intended to squat. They squatted throughout history. With this posture the abdominal wall and bowel are supported as we bear down. This is nature's way.

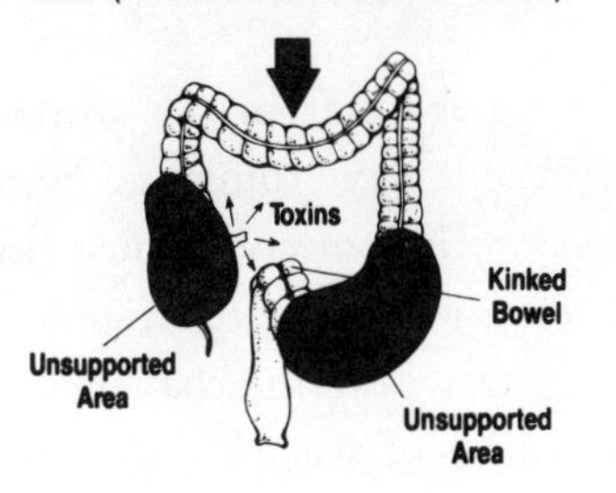

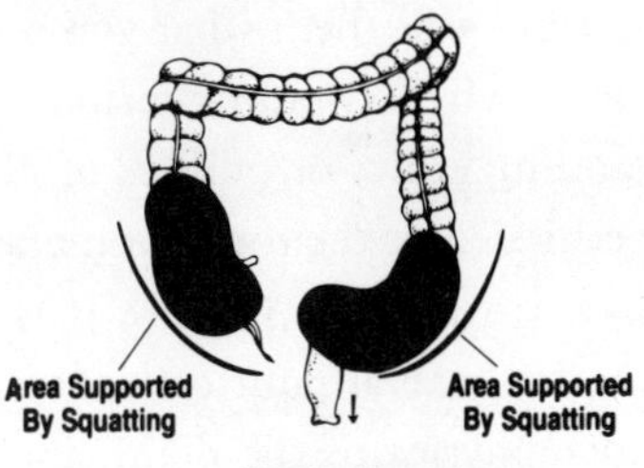

Complete Evacuation

"Western toilet" in his brief account, although one would think that reference to this history might add to the overall critique of "Western" technology. It would also add a dimension that would take it out of the orbit of a critique founded on appeals to "nature's way." Nature's way does not include the social; if it did, the philosophical opposition upon which holistic thought depends would collapse. If nature's authority were to rest upon such a social critique of science and technology, then nature's own "technologies" might be subject to the same critical scrutiny, exposing their own ideological underpinnings. In this respect, the alternative, holistic concept of nature mirrors modern science's desperate attempts to conceal the "hidden crimes" of its own social costs.

In any final analysis, the holistic opposition between "nature" and "technology" is quite groundless. Technologies are social processes of organization, in one form or another, and the sacralized passing of a healer's hands over the body of a patient is just as much a technological process as the demonized use of external hardware like catscans or electroencephalograph machines. For all its philosophical coherence, the New Age community's imposition of *limits* to the categories of nature and technology may be as arbitrary as the process by which the gatekeepers of science go about demarcating science from pseudoscience. But "arbitrary" is obviously not the right word here. The demarcation line that New Age draws between "good" holistic nature and "bad" scientific culture is far from arbitrary, because it always reflects a particular state of power relations. Just as the demarcation line, drawn from above, between science and pseudoscience reflects vested interests on the part of powerful institutions in science, so too the line, drawn from below, between "nature" and "technology" expresses a critique, however mediated and paranoid, of technologism on the part of disempowered nonexperts, scientific autodidacts, technological free spirits—call them what you will.

This is why I would caution against debating, in the abstract, the freestanding ethics or politics of New Age holistic thinking. Even when these ethics enter the realm of metaphysics, their meaning must always be *structurally* evaluated, because it is determined by the relation between a local and a general political context—in this case, the challenge of New Age metaphysics to the dominant certitudes of scientific orthodoxy at a particular moment in the development of Western technocracy.

DISPLACING THE DENTIST

The story I told earlier about my dentist's pseudo-holistic humor hints at the steadily increasing penetration of the medical establishment by the ideas and values of holistic health culture. By the late eighties, holistic values had at least one foot in the door of many major institutional centers of Western medicine. For many healers, this development is potentially fatal, heralding the absorption of a minority culture that had hitherto enjoyed its own non-institutionalized freedoms. Others, more enthusiastic about the prospect of a semi-professionalized challenge to the mechanistic foundations of biomedicine, could see that they would have to argue at closer quarters if the extra-somatic factors pertaining to holistic, as well as environmental and psychosocial, medicine are to transform the dominant picture of disease causation. Paradigmatic treatment of the diseased body would have to include the concept of the environmentally "sick person." In addition, then, to the challenge of the insurgents' *alternative* model of health, a more effective contest would be needed in the form of a thoroughgoing foundational critique.

Laurence Foss and Kenneth Rothenberg, authors of *The Second Medical Revolution*, argue that the insurgent movements have been unable to mount such a critique because their models defer too much to the presuppositions of biomedicine; alternative health increasingly has had to speak the same language of disease, cure, and cause as the dominant paradigm.[34] Maverick clinical successes are not powerful enough; what is needed is a new explanatory framework based upon different philosophical premises. Foss and Rothenberg suggest a new foundational paradigm—"infomedicine"—which draws upon postmodern scientific principles grounded in developments in quantum physics, irreversible thermodynamics, and information theory, in which the body is seen as a self-organizing system, open to environmental feedback, both positive and negative. The action that results from an external stimulus is only indirectly linked to the stimulus's effect on a single part of the system; rather, the action is the result of interaction among a whole pattern of relations within the system. Seen as a self-differentiating system within the environment, the patient's entire biocultural identity is thereby affected by local environmental stimuli. In medical terms, Foss and Rothenberg's paradigm means that a disease cure from biomedicine ought to be seen in conjuction with a social cure from environmental medicine and a patient cure from holistic medicine.

This is one example of the way in which holistic conceptions of nature are being used in a rationalist way to contest the dominant rationalist model. On another front, the relaxing effect of the quantum revolution on science's older certitudes about nature has also muddied the waters of rationalism's exclusion of the non-rational. As a result, proponents of metaphysics have enjoyed more legitimacy in recent years than in recent centuries to be able to ask the question: Why is science better as a form of organized knowledge about the object world than metaphysical explanations of the world that appear to be more comprehensive? Or, to put it another way, how can metaphysical theories and explanations taken seriously by millions be ignored or excluded by a small group of powerful people called "scientists"? It is perhaps easier to answer those questions than it is to explain the immense popularity of the "pseudosciences." To address the question of popularity one would have to start by pointing to the communitarian appeal of cultural practices in which anyone, especially the socially powerless, can join, whereas only professional scientists "do" science. One would also have to account for people's need for self-dignity, consistently thwarted by authority in the command environments of their social lives, and consequently translated into a hunger for completion and a recognition of otherness that is outlawed or proscribed by the official version of common sense. One would then have to consider, not just the new metaphysical critique of scientific authority, but also the rejection of the authority of institutionalized religion that makes passive subjects out of its adherents. And that would just be a beginning.

The answer to the first set of questions lies in the status of science as a discourse of power. Here, one might start from Bruno Latour's observation that "irrationality" is always an accusation made by someone who wants someone else out of the way. The consequences of this observation go well beyond the holy war on irrationality waged, say, by CSICOP, who see their crusade as the first and last defense against a resurgence of the Dark Ages. Latour's work exposes the full history of contingent social conditions and processes of persuasion, involving many different actors, by which technology is developed and then, only in the last instance, legitimated in the name of science. His complex analysis of "science in the making" takes us away from abstract debates about the solidity of scientific rationality and into the messy, disorderly world of social relations where what is known under the stable nomenclature of "science" and "technology" is actually the ceaselessly contested result of the power of institutions, the battlefield of rhetoric, the recruitment drives waged to lend authority to theories, the indeterminacy of "facts," hard and soft, the local availability of capital and technical resources, the vagaries of laboratory politics, and the ethnocentrism of theoretical formalism:

> [Science and technology are] a figment of our imagination, or, more properly speaking, the *outcome* of attributing the whole responsibility for producing facts to a happy few. . . . When one accepts the notion of "science and technology," one accepts a package made by a few scientists to settle responsibilities, to exclude the work of the outsiders, and to keep a few leaders.[35]

"Science" is not only the favored words of a few accredited experts, adroit in the art of rhetoric, trying to keep competitors out. It is also the result of a complex attempt to maneuver and position propositions into alignment with powerful interests. This is why foundational arguments for redefining the paradigm, of the kind discussed by Foss and Rothenberg above, are rather weak if they are waged on philosophical grounds alone. To buttress those arguments, one needs the backing of expensive laboratories and research hospitals, the complicity of insurance companies, and much more besides in the way of infrastructural support. So too, the philosophical arguments draw upon scientific theories that have themselves won legitimacy by the same contested procedures of the sort described above. Even weaker are holistic appeals to nature. As Latour points out, when "nature" itself is in dispute, nature cannot be called upon to referee the contest.

To anyone versed in poststructuralist cultural theory, many of Latour's observations will be all too familiar. Adepts of critical theory will recognize the endless signifying chain of textuality, as well as the power of elite institutions to provide authority for magisterial critics' names and the theories to which they subscribe. Today, appeals to "culture" or "value" are no less neutral and every bit as suspect for the cultural critic as appeals to "nature" or "objectivity" are for the scientist. Scientific and technical papers are saturated with familiar narratives that literary theorists have learned to deconstruct as a matter of course. And the theoretical bottom line is that no medium, whether linguistic (the scientific paper), mathematical (the graph), or technological (the microscope, or computer modeling), is a transparent, objective means for communicating meaning or facts.

We cannot rest easy with the relativism that proceeds from this critique, however. It is a perspective that is usually only available to professional critics, of science and culture alike, who have the cultural capital to demystify the power of their own professional authority. The problem of what Latour calls the Great Divide, between bona fide scientific *knowledge* and what it designates as pseudoscientific *belief*, is as politically significant as the battles of legitimation fought by cultural critics over popular and minority cultures; battles often fought, however ingenuously, in the name of those who are represented in and by these cultures. As I argued earlier, the game of drawing the demarcation line is not simply the mark of a recognition that two cultures, with different beliefs or knowledge claims, are unlikely to have much to say to each other on whatever common ground they might share. A great deal of cultural power is mobilized around that division—the power of a dominant culture over a subordinate one—and the lives of a great number of people are lived in the region that harbors the Great Divide.

The divide is usually characterized by knowledge claims seen as universally true, on one side, and local beliefs seen as culturally specific, on the other. The ethnocentricism of this division is quite evident when it applies to the folk sciences of "traditional" cultures, and is part of the pernicious story of Western scientific imperialism. No less prejudicial, however, is the formidable system of exclusionary languages, gestures and judgments that serve to police the borders of rationality in the everyday life of an advanced technocratic society. Applied sociological analysis (of the sort exemplified by Latour) that traces the recruitment networks and pathways of responsibility within a capital-intensive scientific community can help to expose the interests of power that depend upon this policework. To see how this policing is expressed and worked out in the everyday life of nonexperts, a deeper and messier kind of cultural criticism is needed.

Studies of this sort are needed to make sense of the contradictory personal value systems that sociology founders upon.

Scientists, like cultural intellectuals, have to share their lives with nonexperts who *both respect and resent* the experts' authority to judge the rationality of popular representations of science. Those few scientists, like Carl Sagan, whom the media designate as popularizing spokespersons, speak with the conviction that if all science were properly explained to a nonexpert audience, there would be no "pseudoscience." No Gresham's Law of science in which "good" science is driven out by "bad"—in the same way that advocates of high culture used to maintain that a cultural diet that was "good for the people" would dispense with the hunger for "trash" culture.[36] This position is quite naive, not only because it assumes that all truly scientific knowledge claims are ideology-free, but also because it assumes that a body of knowledge claims will continue to make the same non-contextual sense once it is lifted out of its life-support mechanisms—the vast apparatus of life-sustaining tubes, feeds, drips and interfaces that comprises the context of legitimate expertise.

Consider the opening of Sagan's acceptance speech on receiving CSICOP's "In Praise of Reason" award:

> What is skepticism? It's nothing very esoteric. We encounter it every day. When we buy a used car, if we are the least bit wise we will exert some residual skeptical powers—whatever our education has left to us. . . . There is at least a small degree of interpersonal confrontation involved in the purchase of a used car and nobody claims it is especially pleasant. But there is a good reason for it—because if you don't exercise some minimal skepticism, if you have an absolutely untrammeled credulity, there is probably some price you will have to pay later. Then you'll wish you had made a small investment of skepticism early. Now this is not something that you have to go through four years of graduate school to understand. Everybody understands this.[37]

Despite Sagan's insistence that skepticism is nothing high falutin', his frames of reference here all appeal in some way to the order of accumulating cultural capital: "our residual skeptical powers—whatever our education has left to us," "four years of graduate school," and the governing rhetoric of "investment" in an education that will pay off. The title of his talk, "The

Burden of Skepticism," is even more suggestive, since it invokes the expert's version of the "white man's burden"—in this case, the burden of scientists to exercise skepticism on behalf of others. His speech is a good example of how the "rationality" of experts trails its own context along with it into situations where a broad, nonexpert audience is being addressed.

The contradictions of New Age culture offer another perspective on the parochalism of this position. It is clear that the orthodoxy's denigration of New Age practices is in large part a response to an essentially nonexpert constituency that audaciously proclaims its own authority as a culture based on alternative ideas about science and technology. Legitimate scientists pour scorn on the tendency of New Agers to exploit "authority transfer" by cravenly citing the names of legitimate authorities, especially that of Nobel Prize winners, while being "unwilling to abide by the rules by which the scientific community earned its authority in the first place."[38] Consequently, New Age science is seen as a bricolage of pilfered claims from authority, recombined in a zesty pastiche that is all the more appealing to its audience if it can be shown to have incurred the wrath of the establishment. There is no doubt that this kind of piracy of the expert's "property" is sanctioned by anti-authoritarian populism. It is important, however, to see how this bricolage is constructed, however pretentiously, as an attempt to make "common sense" for an audience that both resents its patronization by legitimate authorities, and at the same time feels it is entitled to a higher version of "common sense" than that offered in popular narratives about science that are more unquestioningly celebratory.

More vitriolic yet is the response, offered by humanistic psychologists like Maureen O'Hara (in a field where New Age thinking has made huge inroads), that sees a collectivist, cryptofascist conspiracy in the New Age assault on scientific authority.[39] This is a rather literalist response to a conceit first presented in Marilyn Ferguson's canonical 1980 book, *The Aquarian Conspiracy*, which announced that a large cross-section of the population, from burned-out 1960s radicals to button-down establishment figures, were all coming together in a "conspiratorial" fashion to consummate their newfound common interests in the politics of consciousness. O'Hara's view is also a reactionary response (similar in form to the "fear of anarchy" that preserves the rule of centralized state power) to the idea that the accumulated consciousness of a significant mass of people, however disparate in social composition, can actually transform cultural reality, let alone "make history," on the basis of their collective will alone.

On the other hand, O'Hara's reaction might appear more justified in the case of certain fringe New Age practices—belief in the ability to usher in millennial Harmonic Convergences, for example. But even this misses the point. Spectacles like the Harmonic Convergence are not pragmatic attempts to realize a philosophical, let alone a political, goal. They are ritual expressions, necessarily utopian, of the cultural desires of a particular community, similar in many ways to the ritualized, public spectacles of NASA launches; both are theatrical exercises in faith whose presentation is governed by their own logically consistent rules. The pragmatic "success" of most NASA launches really only demonstrates science's capacity to verify, by political as well as technical means, its own self-defined successes—an increasingly risky premise in the wake of the Challenger disaster, the recent string of failures in the shuttle program, and the Hubble telescope fiasco. By that same token, the non-verifiable "failure" of the Harmonic Convergence only affirms that the millenarian faith of its audience in a historic transformation must, like all utopian schemes, always be deferred (the latest date, I believe, is 2012) into the realm of the "not yet."[40]

NEW AGE POLITICS

It is now time to draw together some of the political observations I have made about New Age culture in the course of this chapter. Just as I have insisted that New Age challenges to scientific rationality bear the impress of legitimate structures of rationality, however transformed, so too it would be unwise to evaluate the politics of the New Age movement in the abstract, as a freestanding platform with an autonomous viewpoint on society. Just as one is likely to find in New Age culture some of the best and the worst ideas about science and technology, the same configuration holds true for the political ideas to be found in the community. My own affiliations are to cultural and political communities that do not always respond sympathetically to New Age cultural practices, and so I will only propose to represent the claims of New Age politics inasmuch as they speak to a political (mostly socialist) sensibility that, in the New Age mind, has long been superseded but which ought, in my opinion, to be able to extend an interested critique.

By far the majority of New Agers are economically comfortable white folks; one prominent publication, the *Llewellyn Times*, claims that 70 per cent of New Agers are of "Celtic" descent. But there is no tidy class constituency to speak of; audiences range from office workers to business executives, and from technicians to administrators. Prominent New Age ideologues are nonetheless inclined to present the whole community as a decentralized avant garde with a theory of global human transformations, and with the desire, like all vanguards, to initiate people into the cause of making a radical break with history. Also like all vanguards, the responsibility for the Aquarian revolution lies in a creative minority, while the conditions for a revolutionary transformation require only a sufficient number of consciousness-changing individuals. As Ferguson puts it, this minority is "a conspiracy without a political doctrine. Without a manifesto. With conspirators who seek power only to disperse it. . . . Broader than reform, deeper than revolution, this benign conspiracy for a new human agenda has triggered the most rapid cultural realignment in history."[41] As a vanguard group, their inability to hasten the fulfillment of their goals is often rationalized as "popular" resistance to change on the part of conservative, change-fearing masses. On the other hand, New Age recruitism is tempered by the native tradition of appeals to voluntarism, a transformational appeal that the US Army—"Be All That You Can Be"—has worked to a fine point in recent years.

Despite the appeals to the history of American transcendentalism and its various great awakenings, most of the transformational rhetoric of New Age comes from the 1960s counterculture, and it is there that any narrative of the history of New Age politics is obliged to begin. There already exists a dominant media narrative about "the sixties" that involves New Age and that recounts the falling off of radicalism, and the absorption, recuperation, commodification or Yuppification of countercultural politics. It is a narrative about how promises of postscarcity that once encouraged a glut of cultural transgressions now sanction the culture of Yuppie gluttony; how a liberatory revolution in individual rights became a privatized cult of self-interest; how the feminist watchword, "the personal is the political," was inverted to become "the political is the personal"; how the bohemian rejection of materialism became an acceptance of corporate transmaterialism; how the radical otherness of Eastern spirituality was distilled into a radical Protestantism, fully aligned with the American work ethic; how the medium of personal transformation evolved from outlawed psychedelia to corporate high-technology; how the collectivist experiments in cooperative community living became experiments in small-scale entrepreneurship; how radical libertarianism became neoliberalism; how the concept of consciousness-raising groups became a seminar model for executive corporate philosophy; how the hunger for collectivism was taken up in the concept of social networking, a parody of the idea of participatory democracy; how respect for E.F. Schumacher's small-scale economics turned into enthusiasm for John Naisbitt's "megatrends"; how Blakean innocence became social mystification; and how the semiotic gestures of disaffiliation from a dominant culture became incorporated in the business of marketing lifestyles.

This is by now a familar story, told by the media and at least partially accepted, it appears, by large sectors of society. It is often accompanied by the testimony of prominent New Left apostates, among whom Jerry Rubin usually figures as the most likely suspect to be quoted as favoring changes in consciousness over changes in social structure. It is presented as a story about the maturing of a generation, who have inevitably made their peace with capitalism. But it is also a narrative favored by strong patrician voices on the left, where the tradition of cultural despair and pessimism runs deep, and where the burden of that tradition weighs too heavily upon younger generations. For the left, it is a narrative of decline; a narrative we cannot afford, since its only reward lies in prizing the past, or, alternatively, in seeing the "excesses" of sixties "irrationalism" as an infantile disorder that led inevitably to excesses of self-interest in the decades that followed—similar in form, then, to the conservative "I told you so" narrative. The conventional antidotes offered by this narrative are, respectively, collectively mobilizing nostalgia; or a responsible patrician call for self-restraint, deference to public interest, and heroic affirmation of traditional reason. Neither antidote speaks very persuasively to the changed cultural landscape since the 1960s, or to the rise of new utopian social movements centered around ecology, gender, minority rights, and sexual preference. Above all, neither speaks to the shortcomings in leftist thinking that New Left subjectivism sought to address—the failure of collectivist thinking to account for the importance of "private" or personal acts in people's everyday lives, and, consequently, the failure affectively to persuade people to abandon an individualist for a collectivist identity.

The rise of Thatcherism and Reaganism was due, in large part, to the New Right's ability to give reactionary form and shape to the focus on individual responsibility that emerged from the crisis of political paternalism in the 1960s. Appeals to individual consumerism and guarantees of the autonomous right to choose were successful in channeling and containing the desires for individual citizenship that were liberated under the concept of participatory democracy. In the cultural vacuum created by the absence of a leftist alternative to progressive individualism, alternative cultures such as the New Age movement evolved in the 1970s. New Age took to heart the sixties languages of personal and everyday life transformation, grafting them on to the grammar of self-determination that eschewed the materialism of acquisitive self-interest espoused by the New Right, while offering a weak vocabulary of social responsibility somewhat removed from the politics of class, race, gender, and sexual preference waged by the post-sixties left.

It is in New Age, then, that we can find at least one version, however troubled, and however incipiently conservative, of a politics of identity and subjectivity that has not been entirely subsumed by the New Right's picture of individualism, aggressively protective of its privatized will to consume. Nor can this culture of self-development be entirely collapsed into the orgiastic picture of "narcissism" offered by conservative left patricians like Christopher Lasch, Daniel Bell, and Russell Jacoby; or into the class "error" of petty-bourgeois self-interest offered by the unreconstructed left. If New Age politics exists, then it does start by taking politics personally, often as a crucible for "born again" ritualism. Any secular end to this process of personal transformation depends upon the naive assumption that changing the self will change the world, an equation that might point to the lack of available languages for linking subjectivity with larger social or structural change. If, on the other hand, the end result of New Age exercises in personal transformation invariably lie in mysticism, then we are obliged to consider how and why the personal desires set in motion there cannot be consummated in everyday social life. Without any alternative (left) politics of social individualism it is no surprise to find that people will opt for spiritualist solutions for the problems encountered in a materialistic culture. This is surely one of the lessons to be drawn from the worldwide revival of religious fundamentalism, as well as the flood of non-institutional cults, old and new, that have poured into the respective vacuums created by the bankrupt cultures of state socialist, capitalist, and scientific materialism.

In this respect, New Age politics might be seen not as a wayward, pathological creature of the New Left's imagination, but as a political innocent in candid, questioning dialogue with the unclaimed mainstream territory of progressive, rather than atomistic, individualism. Indeed, if we were to examine some of the social and political threads that run through the aery fabric of New Age thinking, we would find certain themes that resonate with the necessary conditions for a left version of progressive individualism. Generally speaking, New Age addresses its adherents as active participants, with a measure of control over their everyday lives, and not as passive subjects, even victims, of larger, objective forces. The New Age "person" is also in many respects an individual whose personal growth is indissociable from the environment; a link fleshed out in a variety of ecotopian stories and romances. So, too, the small-scale imperative of New Age's cooperative communitarianism brings with it a host of potentially critical positions: against big, centralized bureaucracies; against big, transnational business conglomerates; against large-scale, and environmentally destructive, technologies; against the imperialist claims made on the basis of strong nationalism; and against monolithic institutions in education, industry, religion, and the nuclear family. The focus on self-empowerment might be extended to include forms of local self-government and citizen diplomacy, worker self-management, job-sharing, family role-sharing, reproductive self-determination, children's rights, and healthcare rights. Last but not least, the pervasive influence on mainstream consciousness of the dietary and nutritional emphasis of holistic thought and practice has had an enormous impact on public health.[42]

Above all, New Age consciousness, whose activist roots lies in a deep, mystical affinity with nature, has played an important role in shaping the social and cultural activism of the ecology movement, increasingly divided between the philosophies of deep ecology and social ecology. Neither side, however, seems to want to claim New Age as an official ally. Deep ecologists denigrate New Age's evolutionary devotions to growth as a technocratic strategy for using the earth as an expendable resource, citing Buckminster Fuller's image of "Spaceship Earth" as an example of New Agers' technohumanist contempt for a planet that, once exhausted, will then be left behind.[43] Social ecologists, on the other hand, equate New Age influence with the mystical nature cults, wiccan goddesses and all, that have saturated the eco-feminist and deep ecology constituencies of the movement with an atavistic taste for supernaturalism. From this perspective, New Age thought is seen as a dangerously anti-social element, exhausting the rationalist reserves needed to reconstruct a free society living in non-dominating balance with the natural world.

There are lessons to be drawn from both of these critiques. On the one hand, the commitment of New Age humanism to unlimited personal growth, by whatever appropriate technological means, excludes the more organic view of coexistence with the natural world. I find the latter critique more persuasive, however. New Age shares deep ecology's championing of voluntary simplicity and its religious opposition between "nature" and "society/technology." This construction of nature as a social vacuum distances us from any direct engagement with the actual social forces that command vast power in our everyday lives through their organization of technology and bureaucracy. One of the inevitable effects of this retreat is to entertain Arcadian fantasies of preindustrialist life resourcefully embellished with many of the philosophical contents of a postindustrialist wardrobe. To advocate a reformist program of massively cleansing the technomaterial world in the interests of social purity is to turn a blind eye to the enormous human costs involved in, for example, deindustrialization, especially in the world's developing countries. Conceived as a social program, this would require a large measure of equal access to technology and thus of control over the development of more democratic, ecologically sound technologies.

Similar critiques apply to New Age individualism. When personal consciousness is the single determining factor in social change, then all social problems, including the specters raised by racism, imperialism, sexism, and homophobia, are seen as the result of personal failures and shortcomings. Individual consciousness becomes the source, rather than a major site, of socially oppressive structures, and the opportunities for a radical humanism are consequently lost. A radical politics of personalism stands to learn a good deal from the deeply felt response of New Age humanism to large-scale technological organization, especially those organizations of political rationality whose explanatory social models exclude the politics of everyday life and subjectivity. But a radical humanism can hardly afford to jettison its social critiques of the institutions and ideologies of unfreedom in favor of an individual voluntarism that quarantines the human or individual from the social.

The combined force of these critiques ought to serve as a further caution against seeing "New Age politics" as a freestanding platform. But there remains the obligation to try to understand how and why the limitations of New Age politics are set in the ways that I have tried to describe. One of these explanations, as I have suggested, lies in the absence of a credibly persuasive left alternative that might draw out and extend into a coherent philosophy of social individualism those features of New Age personalism that arise from a critical response to a technocratic society. The absence of that contesting influence has surely left the field open to the powerful influence of a more conservative individualism that defines personal self-interest as a realm around which all social action can be thought and pursued.

But this view of contesting forces is too cut and dried, resting upon a rather patronizing code of recruitism, whereby errant members of the flock can be retrieved simply by changing the pitch and tone of the shepherd's call. Such an explanation undervalues, if it does not altogether ignore, the powerful desire for self-respect, self-determination, and utopian experimentalism that lies behind the development of an alternative cultural community like the New Age movement. This desire, inspired by a deep hunger for community, provides such a culture with a coherent *raison d'être* in the first place, and surely saves it from being a disorganized set of psychosocial impulses. It is this desire that enables the New Age movement to present a relatively coherent challenge to medical, scientific and religious orthodoxies. And yet it is the same desire for self-coherence that distances New Age thinking from thoroughgoing social critiques of the institutions that house these same orthodoxies. Therein lie both the fate and the strength of self-coherent countercultures that try to live out aspects of utopian futures in the present, making histories, as always, under conditions not entirely of their own making.

One of the undeniable strengths of cultural studies has lain in its willingness to explain the political significance of such countercultures. Their practices offer less articulate, less pure, and less overtly political kinds of cultural critique than the left has traditionally felt comfortable endorsing. More articulate critiques of science than those offered by New Age certainly do exist and flourish: in medicine, the still thriving alternatives offered by the women's healthcare movement, motivated by feminist critiques of the history of medical science; and, most recently, the struggle of PWAs and others in the AIDS community to redefine the praxis of medical research by using their own bodies as socialized laboratories. I am sure that most readers could cite any number of activist movements around issues of science and technology: nuclear technologies, biotechnologies, pesticides, toxic biohazards, animal rights, and so on. When it comes to political evaluation of New Age culture, we are on less secure ground and therefore miss some of the more familiar landmarks for recognizing social and political criticism. At the beginning of a decade that will surely witness a whole range of political appeals to different notions of "nature," and different interpretations of (limits to) "growth," New Age definitions of these categories can serve both as a caveat to and a stimulus for our efforts to shape a green cultural criticism and politics.

In spite of its sheer diversity and apparent tolerance for all manner of esoteric practices and disciplines, I have nevertheless assumed in this chapter that New Age culture is the product of a highly structured community of interests, strengthened by the moralistic nature of its vanguardist resolve. In the absence of any central institutional forum, its networking communities and the internal debates about the direction of the New Age movement (quite explicit in the pages of prominent magazines like *New Age*) take on the function of regulating the codes that hold the disparate range of practices and disciplines together. I have tried to describe some of these codes, in particular those that have a bearing on the science/pseudoscience divide, on the one hand, and those that serve to define "external" technologies as inappropriate, on the other. In the former case, I argued that New Age "rationality" was pervasively characterized by the simultaneous activities of emulating, pirating, and rejecting the authority of orthodox scientific rationality. In the latter case, I examined some of the unanswered social questions that lie behind the naturalist philosophies of the psychotechnologies. Both cases, however, involve a challenge to the process by which demarcation lines are legitimately drawn. In its embrace of the new science, New Age encroaches upon and confuses the delimiting tasks of science's traditional bordermakers and gatekeepers. In the case of appropriate technology, the limit is self-imposed from within the New Age community itself, and is therefore a code that appropriates from science the power of the experts to draw up the borderlines.

I want to conclude by suggesting that New Age culture can also be seen as a particular version of the arguments raised in the debates about postmodernism. For in New Age we can see an expression of one of our most urgent political horizons—the need to be able to go on speaking about growth—human and/or social—while recognizing the need to acknowledge ecological limits to technological growth and development. In this respect, New Age is a response, if you like, to the so-called Enlightenment "project of modernity" that was, and still is, bound up with the imperatives of growth and development. In principle, New Age proposes a continuation of this project, but in the name of a different human rationality. This proposal has something in common with social theorists like Jürgen Habermas, who speaks of the "incomplete project of modernity," and who wants to renew the foundational principles of Enlightenment rationality, ever distorted by capitalist forms of instrumental reason and technorationality.[44] New Age ideas of "rationality" may be as distant from Habermas's ideal of communicative rationality as they are from poststructuralists' emphasis on "difference," but each is a dissenting response to dominant forms of Western rationality; each is a symptom of the crisis of technical rationality that "postmodernism," in its many forms, has come to address.

The most crucial aspect of the postmodernism debate, and perhaps the least discussed, is its bearing upon the ecological questions of growth and development. In this respect, there may be a concluding lesson to be drawn from the New Age movement. The New Age community itself comprises a sizable population who believe that there ought to exist limits to the authority of science and to technological growth. The community shares a fiercely moral attachment to the idea of setting external limits, while at the same time being devoted to limitless "internal" development. What we must remember, however, especially in the years of green politics and activism ahead, is that there is a difference between saying that limits *ought* to exist, and saying that we ought to recognize the existence of limits. The first, the New Age formulation, rests upon an absolute philosophical imperative, and appeals to an absolute and thoroughly unsocialized view of nature. The second is the formulation that I recommend; it rests its case on the evidence that limits to growth may be socially desirable, and can be shown to be so. Such limits cannot be dictated; they must be freely chosen, without falling back on appeals to nature's perennial wisdom, and without forsaking the differentiated goals of social growth for the purely individualistic or universal goals of human growth.

CHAPTER TWO

HACKING AWAY AT THE COUNTERCULTURE

Ever since the viral attack engineered in November 1988 by Cornell University hacker Robert Morris on the national network system Internet, which includes the Pentagon's ARPAnet data exchange network, the nation's high-tech ideologues and spin doctors have been locked in debate, trying to make ethical and economic sense of the event. The virus rapidly infected an estimated six thousand computers around the country, creating a scare that crowned an open season of viral hysteria in the media, in the course of which, according to the Computer Virus Industry Association in Santa Clara, California, the number of known viruses jumped from seven to thirty during 1988, and from three thousand infections in the first two months of that year to thirty thousand in the last two months.[1] While it caused little in the way of data damage (some richly inflated initial estimates reckoned up to $100 million in downtime), the ramifications of the Internet virus have helped to generate a moral panic that has all but transformed everyday "computer culture."

Following the lead of the Defence Advance Research Projects Agency (DARPA) Computer Emergency Response Team at Carnegie-Mellon University, antivirus response centers were hastily put in place by government and defense agencies at the National Science Foundation, the Energy Department, NASA and other sites. Plans were made to introduce a bill in Congress (the Computer Virus Eradication Act, to replace the 1986 Computer Fraud and Abuse Act, which pertained solely to government information) that would call for prison sentences of up to ten years for the "crime" of sophisticated hacking, and numerous government agencies have been involved in a proprietary fight over the creation of a proposed Center for Virus Control—modeled, of course, on Atlanta's Centers for Disease Control, notorious for its failures to respond adequately to the AIDS crisis.

In fact, media commentary on the virus scare has run not so much tongue-in-cheek as hand-in-glove with the rhetoric of AIDS hysteria—for example, the common use of terms like *killer virus* and *epidemic*; the focus on high-risk personal contact (virus infection, for the most part, is spread through personal computers, not mainframes); the obsession with defense, security, and immunity; and the climate of suspicion generated around communitarian acts of sharing. The underlying moral imperative is this: you can't trust your best friend's software any more than you can trust his or her bodily fluids. Safe software or no software at all! Or, as Dennis Miller put it on *Saturday Night Live*, "Remember, when you connect with another computer, you're connecting to every computer that computer has ever connected to." This playful conceit struck a chord in the popular consciousness, even as it was perpetuated in such sober quarters as the Association for Computing Machinery—the president of which, in a controversial editorial titled "A Hygiene Lesson," drew comparisons not only with sexually transmitted diseases but also with a cholera epidemic, and urged attention to "personal systems hygiene."[2] Some computer scientists who studied the symptomatic path of Morris's virus across Internet have pointed to its uneven effects upon different computer types and operating systems, and concluded that "there is a direct analogy with biological genetic diversity to be made."[3] The epidemiology of biological virus (especially AIDS) research is being studied closely to help implement computer security plans. In these circles, the new witty discourse is laced with references to antigens, white bloodcells, vaccinations, metabolic free radicals, and the like.

The form and content of more lurid articles like *Time*'s infamous story, "Invasion of the Data Snatchers" (September 1988), fully displayed the continuity of the media scare with those historical fears about bodily invasion, individual and national, that are endemic to the paranoid style of American political culture.[4] Indeed, the rhetoric of computer culture, in common with the medical discourse of AIDS research, has fallen in line with the paranoid, strategic mode of Defense Department rhetoric established during the Cold War. Each language repertoire is obsessed with hostile threats to bodily and technological immune systems; every event is a ballistic maneuver in the game of microbiological war, where the governing metaphors are indiscriminately drawn from cellular genetics and cybernetics alike. As a counterpoint to the tongue-in-cheek AI tradition of seeing humans as "information-exchanging environments," the imagined life of computers has taken on an organicist shape now that they too are subject to cybernetic "sickness" or disease. So too, the development of interrelated systems, such as Internet itself, has further added to the structural picture of an interdependent organism, whose component members, however autonomous, are all nonetheless affected by the "health" of each individual constituent. The growing interest among scientists in developing computer programs that will simulate the genetic behavior of living organisms (in which binary numbers act like genes) points to a future where the border between organic and artificial life is less and less distinct.

In keeping with the increasing use of the language of biology to describe mutations in systems theory, conscious attempts to link the Aids crisis with the information security crisis have pointed out that both kinds of virus, biological and electronic, take over the host cell/program and clone their carrier genetic codes by instructing the hosts to make replicas of the viruses. Neither kind of virus, however, can replicate itself independently; both are pieces of code that attach themselves to other cells/programs—just as biological viruses need a host cell, computer viruses require a host program to activate them. The Internet virus was not, in fact, a virus, but a worm, a program that can run independently and therefore appears to have a life of its own. The worm replicates a full version of itself in programs and systems as it moves from one to another, masquerading as a legitimate user by guessing the user passwords of locked accounts. Because of this autonomous existence, the worm can be seen to behave as if it were an organism with some purpose, or teleology, and yet it has none. Its only "purpose" is to reproduce and infect. If the worm has no inbuilt anti-replication code, or if the code is faulty as was the case with the Internet worm, it will make already infected computers repeatedly accept further replicas of itself until their memories are clogged. A much quieter worm than that engineered by Morris would have moved more slowly (as one supposes a "worm" should), protecting itself from detection by ever more subtle camouflage, and propagating its cumulative effect of operative systems inertia over a much longer period of time.

In offering such descriptions, however, we must be wary of attributing a teleology/intentionality to worms and viruses that can be ascribed only, and in most instances speculatively, to their authors. There is no reason why a cybernetic "worm" might be expected to behave like a biological worm in any fundamental way. So, too, the assumed intentionality of its author distinguishes the human-made cybernetic virus from the case of the biological virus, the effects of which are fated to be received and discussed in a language saturated with human-made structures and narratives of meaning and teleological purpose. Writing about the folkloric theologies of significance and explanatory justice (usually involving retribution) that have sprung up around the AIDS crisis, Judith Williamson has pointed to the implications of this collision between an intentionless virus and a meaning-filled culture:

> Nothing could be more meaningless than a virus. It has no point, no purpose, no plan; it is part of no scheme, carries no inherent significance. And yet nothing is harder for us to confront than the complete absence of meaning. By its very definition, meaninglessness cannot be articulated within our social language, which is a system *of* meaning: impossible to include, as an absence, it is also impossible to exclude—for meaninglessness isn't just the opposite of meaning, it is the end of meaning, and threatens the fragile structures by which we make sense of the world.[5]

No such judgment about meaninglessness applies to the computer security crisis. In contrast to HIV's lack of meaning or intentionality, cybernetic viruses are always replete with social significance. Their meaning is related, first of all, to the author's local intention or motivation—whether psychic or fully social, whether wrought out of a mood of vengeance, a show of bravado or technical expertise, a commitment to a political act, or in anticipation of the profits that often accrue from the victims' need to buy an antidote from the author. Beyond these local intentions, however, which are usually obscure or, as in the Morris case, quite inscrutable, there is an entire set of social and historical narratives that surround and are part of the "meaning" of the virus: the coded anarchist history of the youth hacker subculture; the militaristic environments of search-and-destroy warfare (a virus has two components—a carrier and a "warhead"), which, because of the historical development of computer technology, constitute the family values of information technoculture; the experimental research environments in which creative designers are encouraged to work; and the conflictual history of pure as against applied ethics in the science and technology communities, to name just a few. A similar list could be drawn up to explain the widespread and varied *response* to computer viruses, from the amused concern of the cognoscenti to the hysteria of the casual user, and from the research community and the manufacturing industry to the morally aroused legislature and the mediated culture at large. Every one of these explanations and narratives is the result of social and cultural processes and values; consequently, there is very little about the virus itself that is "meaningless." Viruses can no more be seen as an objective or necessary result of the "objective" development of technological systems than technology in general can be seen as an objective, determining agent of social change.

For the sake of polemical economy, I would note that the cumulative effect of all the viral hysteria has been twofold. Firstly, it has resulted in a windfall for software producers, now that users' blithe disregard for makers' copyright privileges has eroded in the face of the security panic. Used to fighting halfhearted rearguard actions against widespread piracy practices, or reluctantly acceding to buyers' desire for software unencumbered by top-heavy security features, software vendors are now profiting from the new public distrust of program copies. So, too, the explosion in security consciousness has hyper-stimulated the already fast-growing sectors of the security system industry and the data encryption industry. In line with the new imperative for everything from "vaccinated" workstations to "sterilized" networks, it has created a brand new market of viral vaccine vendors who will sell you the virus (a one-time only immunization shot) along with its antidote—with names like Flu Shot +, ViruSafe, Vaccinate, Disk Defender, Certus, Viral Alarm, Antidote, Virus Buster, Gatekeeper, Ongard, and Interferon. Few of the antidotes are very reliable, however, especially since they pose an irresistible intellectual challenge to hackers who can easily rewrite them in the form of ever more powerful viruses. Moreover, most corporate managers of computer systems and networks know that the vast majority of their intentional security losses are a result of insider sabotage and monkeywrenching.

In short, the effects of the viruses have been to profitably clamp down on copyright delinquency, while generating the need for entirely new industrial production of viral suppressors to contain the fallout. In this respect, it is easy to see that the appearance of viruses could hardly, in the long run, have benefited industry producers more. In the same vein, the networks that have been hardest hit by the security squeeze are not restricted-access military or corporate systems but networks like Internet, set up on trust to facilitate the open academic exchange of data, information, and research, and watched over by its sponsor, DARPA. It has not escaped the notice of conspiracy theorists that the military intelligence community, obsessed with "electronic warfare," actually stood to learn a lot from the Internet virus; the virus effectively "pulsed the system," exposing the sociological behavior of the system in a crisis.[6]

The second effect of the virus crisis has been more overtly ideological. Virus-conscious fear and loathing have clearly fed into the paranoid climate of privatization that increasingly defines social identities in the new post-Fordist order. The result—a psychosocial closing of the ranks around fortified private spheres—runs directly counter to the ethic that we might think of as residing at the architectural heart of information technology. In its basic assembly structure, information technology involves processing, copying, replication, and simulation, and therefore does not recognize the concept of private information property. What is now under threat is the rationality of a shareware culture, ushered in as the achievement of the hacker counterculture that pioneered the personal computer revolution in the early 1970s against the grain of corporate planning.

There is another story to tell, however, about the emergence of the virus scare as a profitable ideological moment, and it is the story of how teenage hacking has come to be defined increasingly as a potential threat to normative educational ethics and national security alike. The story of the creation of this "social menace" is central to the ongoing attempts to rewrite property law in order to contain the effects of the new information technologies that, because of their blindness to the copyrighting of intellectual property, have transformed the way in which modern power is exercised and maintained. Consequently, a deviant social class or group has been defined and categorized as "enemies of the state" in order to help rationalize a general law-and-order clampdown on free and open information exchange. Teenage hackers' homes are now habitually raided by sheriffs and FBI agents using strong-arm tactics, and jail sentences are becoming a common punishment. Operation Sun Devil, a nationwide Secret Service operation conducted in the spring of 1990, and involving hundreds of agents in fourteen cities, is the most recently publicized of the hacker raids that have resulted in several arrests and the seizure of thousands of disks and address lists in the last two years.[7]

In one of the many harshly punitive prosecutions against hackers in recent years, a judge went so far as to describe "bulletin boards" as "high-tech street gangs." The editors of *2600*, the magazine that publishes information about system entry and exploration indispensable to the hacking community, have pointed out that any single invasive act, such as trespass, that *involves* the use of computers is considered today to be infinitely more heinous than a similar act undertaken *without* computers.[8] To use computers to execute a prank, raid, fraud or theft is to incur automatically the full repressive wrath of judges, urged on by the moral panic created around hacking feats over the last two decades. Indeed, a strong body of pressure groups is pushing for new criminal legislation that will define "crimes with computers" as a special category deserving "extraordinary" sentences and punitive measures. An increasingly criminal connotation today has displaced the more innocuous, amateur-mischief-maker-cum-media-star role reserved for hackers until a few years ago.

In response to the gathering vigor of this "war on hackers," the most common defenses of hacking can be presented on a spectrum that runs from the appeasement or accommodation of corporate interests to drawing up blueprints for cultural revolution: (a) Hacking performs a benign industrial service of uncovering security deficiencies and design flaws. (b) Hacking, as an experimental, free-form research activity, has been responsible for many of the most progressive developments in software development. (c) Hacking, when not purely recreational, is an elite educational practice that reflects the ways in which the development of high technology has outpaced orthodox forms of institutional education. (d) Hacking is an important form of watchdog counter-response to the use of surveillance technology and data-gathering by the state, and to the increasingly monolithic communications power of giant corporations. (e) Hacking, as guerrilla knowhow, is essential to the task of maintaining fronts of cultural resistance and stocks of oppositional knowledge as a hedge against a technofascist future. With all of these and other arguments in mind, it is easy to see how the social and cultural *management* of hacker activities has become a complex process that involves state policy and legislation at the highest levels. In this respect, the virus scare has become an especially convenient vehicle for obtaining public and popular consent for new legislative measures and new powers of investigation for the FBI.[9]

Consequently, certain celebrity hackers have been quick to play down the zeal with which they pursued their earlier hacking feats, while reinforcing the *deviant* category of "technological hooliganism" reserved by moralizing pundits for "dark-side" hacking. Hugo Cornwall, British author of the bestselling *Hacker's Handbook*, presents a Little England view of the hacker as a harmless fresh-air enthusiast who "visits advanced computers as a polite country rambler might walk across picturesque fields." The owners of these properties are like "farmers who don't mind careful ramblers." Cornwall notes that "lovers of fresh-air walks obey the Country Code, involving such items as closing gates behind one and avoiding damage to crops and livestock" and suggests that a similar code ought to "guide your rambles into other people's computers; the safest thing to do is simply browse, enjoy and learn." By contrast, any rambler who "ventured across a field guarded by barbed wire and dotted with notices warning about the Official Secrets Act would deserve most that happened thereafter."[10] Cornwall's quaint perspective on hacking has a certain "native charm," but some might think that this beguiling picture of patchwork-quilt fields and benign gentlemen farmers glosses over the long bloody history of power exercised through feudal and post-feudal land economy in England, while it is barely suggestive of the new fiefdoms, transnational estates, dependencies, and principalities carved out of today's global information order by vast corporations capable of bypassing the laws and territorial borders of sovereign nation-states. In general, this analogy with "trespass" laws, which compares hacking to breaking and entering other people's homes, restricts the debate to questions about privacy, property, possessive individualism, and, at best, the excesses of state surveillance, while it closes off any examination of corporate owners' and institutional sponsors' information technology activities (the most prized "target" of most hackers).[11]

Cornwall himself has joined the well-paid ranks of ex-hackers who either work for computer security firms or write books about security for the eyes of worried corporate managers.[12] A different, though related, genre is that of the penitent hacker's "confession," produced for an audience thrilled by tales of high-stakes adventure at the keyboard, but written in the form of a computer security handbook. The best example of the "I Was a Teenage Hacker" genre is Bill (aka "The Cracker") Landreth's *Out of the Inner Circle: The True Story of a Computer Intruder Capable of Cracking the Nation's Most Secure Computer Systems*, a book about "people who can't 'just say no' to computers." In full complicity with the deviant picture of the hacker as "public enemy," Landreth recirculates every official and media cliché about subversive conspiratorial groups by recounting the putative exploits of a high-level hackers' guild called the Inner Circle. The author himself is presented in the book as a former keyboard junkie who now praises the law for having made a moral example of him:

> If you are wondering what I am like, I can tell you the same things I told the judge in federal court: Although it may not seem like it, I am pretty much a normal American teenager. I don't drink, smoke or take drugs. I don't steal, assault people, or vandalize property. The only way in which I am really different from most people is in my fascination with the ways and means of learning about computers that don't belong to me.[13]

Sentenced in 1984 to three years' probation, during which time he was obliged to finish his high school education and go to college, Landreth concludes: "I think the sentence is very fair, and I already know what my major will be. . . ." As an aberrant sequel to the book's contrite conclusion, however, Landreth vanished in 1986, violating his probation, only to later face a stiff five-year jail sentence—a sorry victim, no doubt, of the recent crackdown.

CYBER-COUNTERCULTURE?

At the core of Steven Levy's 1984 bestseller, *Hackers*, is the argument that the hacker ethic, first articulated in the 1950s among the famous MIT students who developed multiple-access user systems, is libertarian and crypto-anarchist in its right-to-know principles and its advocacy of decentralized technology. This hacker ethic, which has remained the preserve of a youth culture for the most part, asserts the basic right of users to free access to all information. It is a principled attempt, in other words, to challenge the tendency to use technology to form information elites. Consequently, hacker activities were presented in the 1980s as a romantic countercultural tendency, celebrated by critical journalists like John Markoff of the *New York Times*, and Stewart Brand of *Whole Earth Catalog* fame, and by New Age gurus like Timothy Leary in the magazine *Reality Hackers*. Fueled by sensational stories about "phone phreaks" like Joe Egressia (the blind eight-year-old who discovered the phone company's tone signal by whistling) and Captain Crunch, groups like the Milwaukee 414s, the Los Angeles ARPAnet hackers, the SPAN Data Travelers, the Chaos Computer Club of Hamburg, the British Prestel hackers, *2600*'s BBS, "The Private Sector," and others, the dominant media representation of the hacker came to be that of the "rebel with a modem," to use Markoff's term, at least until the more recent "war on hackers" began to shape media coverage.

On the one hand, this popular folk hero persona offered the romantic high profile of a maverick though nerdy cowboy whose fearless raids upon an impersonal "system" were perceived as a welcome tonic in the gray age of technocratic routine. On the other hand, he was something of a juvenile technodelinquent who hadn't yet learned the difference between right and wrong; a wayward figure whose technical brilliance and proficiency differentiated him from, say, the maladjusted working-class J.D. street-corner boy of the 1950s (hacker mythology, for the most part, has been almost exclusively white, masculine, and middle class). One result of this media profile was a persistent infantilization of the hacker ethic—a way of trivializing its embryonic politics, however finally complicit with dominant technocratic imperatives or with entrepreneurial–libertarian ideology one perceives these politics to be. The second result was to reinforce, in the initial absence of coercive jail sentences, the high educational stakes of training the new technocratic elites to be responsible in their use of technology. Never, the given wisdom goes, has a creative elite of the future been so in need of the virtues of a liberal education steeped in Western ethics!

The full force of this lesson in computer ethics can be found *laid out* in the official Cornell University report on the Robert Morris affair. Members of the university commission set up to investigate the affair make it quite clear in their report that they recognize the student's academic brilliance. His hacking, morever, is described as a "juvenile act" that had no "malicious intent" but that amounted, like plagiarism, the traditional academic heresy, to a dishonest transgression of other users' rights. (In recent years, the privacy movement within the information community-—mounted by liberals to protect civil rights against state gathering of information—has actually been taken up and used as a means of criminalizing hacker activities.) As for the consequences of this juvenile act, the report proposes an analogy that, in comparison with Cornwall's *mature* English country rambler, is thoroughly American, suburban, middle-class and *juvenile*. Unleashing the Internet worm was like "the driving of a golf-cart on a rainy day through most houses in the neighborhood. The driver may have navigated carefully and broken no china, but it should have been obvious to the driver that the mud on the tires would soil the carpets and that the owners would later have to clean up the mess."[14]

In what stands out as a stiff reprimand for his alma mater, the report regrets that Morris was educated in an "ambivalent atmosphere" where he "received no clear guidance" about ethics from "his peers or mentors" (he went to Harvard!). But it reserves its loftiest academic contempt for the press, whose heroizing of hackers has been so irresponsible, in the commission's opinion, as to cause even further damage to the standards of the computing profession; media exaggerations of the courage and technical sophistication of hackers "obscures the far more accomplished work of students who complete their graduate studies without public fanfare," and "who subject their work to the close scrutiny and evaluation of their peers, and not to the interpretations of the popular press."[15] In other words, this was an inside affair, to be assessed and judged by fellow professionals within an institution that reinforces its authority by means of internally self-regulating codes of professional ethics, but rarely addresses its ethical relationship to society as a whole (acceptance of defense grants, and the like). Generally speaking, the report affirms the genteel liberal ideal that professionals should not need laws, rules, procedural guidelines, or fixed guarantees of safe and responsible conduct. Apprentice professionals ought to have acquired a good conscience by osmosis from a liberal education, rather than from some specially prescribed course in ethics and technology.

The widespread attention commanded by the Cornell report (attention from the Association of Computing Machinery, among others) demonstrates the industry's interest in how the academy invokes liberal ethics in order to assist in managing the organization of the new specialized knowledge about information technology. Despite, or perhaps because of, the report's steadfast pledge to the virtues and ideals of a liberal education, it bears all the marks of a legitimation crisis inside (and outside) the academy surrounding the new and all-important category of computer professionalism. The increasingly specialized design knowledge demanded of computer professionals means that codes going beyond the old professionalist separation of mental and practical skills are needed to manage the division that a hacker's functional talents call into question, between a purely mental pursuit and the pragmatic sphere of implementing knowledge in the real world. "Hacking" must then be designated as a strictly *amateur* practice; the tension in hacking between *interestedness* and *disinterestedness* is different from, and deficient in relation to, the proper balance demanded by professionalism. Alternatively, hacking can be seen as the amateur flipside of the professional ideal—a disinterested love in the service of interested parties and institutions. In either case, it serves as an example of professionalism gone wrong, if not very wrong.

In common with the two responses to the virus scare I described earlier—the profitable reaction of the computer industry, and the self-empowering response of the legislature—the Cornell report shows how the academy uses a case like the Morris affair to strengthen its own sense of moral and cultural authority in the sphere of professionalism, particularly through its scornful indifference to and aloofness from the codes and judgments exercised by the media, its diabolical competitor in the field of knowledge. Indeed, for all the trumpeting about excesses of power and disrespect for the law of the land, the revival of ethics in the business and science disciplines of the Ivy League and on Capitol Hill (both awash with ethical fervor in the post-Boesky/Milker and post-Reagan years) is little more than a weak liberal response to working flaws or adaptational lapses in the technocracy's social logic.

To complete the scenario of morality play example-making, however, we must also consider that Morris's father was chief scientist at the National Computer Security Center, the National Security Agency's public effort at safeguarding computer security. A brilliant programmer and code-breaker in his own right, he had testified in Washington in 1983 about the need to deglamorize teenage hacking, comparing it to "stealing a car for the purpose of joyriding." In a further Oedipal irony, Morris Sr may have been one of the inventors, while at Bell Labs in the 1950s, of a computer game involving self-perpetuating programs that were a prototype of today's worms and viruses. Called Darwin, its principles were incorporated in the 1980s into the popular hacker game Core War, in which autonomous "killer" programs fought each other to the death.[16]

With the appearance in the Morris affair of the Pentagon's guardian angel as patricidal object—an implicated if not victimized father—we now have many of the classic components of countercultural, cross-generational conflict. We might consider how and where this scenario differs from the definitive contours of such conflicts that we recognize as having been established in the 1960s; how the Cornell hacker Morris's relation to, say, campus "occupations" today is different from that evoked by the famous image of armed black students emerging from a sit-in on the Cornell campus; how the relation to technological ethics differs from Andrew Kopkind's famous statement, "Morality begins at the end of a gun barrel," which accompanied the publication of the "do-it-yourself Molotov cocktail" design on the cover of a 1968 issue of the *New York Review of Books*; or how hackers' prized potential access to the networks of military systems warfare differs from the prodigious Yippie feat of levitating the Pentagon building. It may be that, like the J.D. rebel without a cause of the fifties, the disaffiliated student dropout of the sixties, and the negationist punk of the seventies, the hacker of the eighties has come to serve as a visible, public example of moral maladjustment, a hegemonic testcase for redefining the dominant ethics in an advanced technocratic society.

What concerns me here, however, are the different conditions that exist today for recognizing countercultural expression and activism. Twenty years later, the technology of hacking and viral guerrilla warfare occupies a similar place in countercultural fantasy as the Molotov cocktail design once did. While such comparisons are not particularly sound, I do think that they conveniently mark a shift in the relation of countercultural activity to technology; a shift in which a software-based technoculture organized around outlawed libertarian principles about free access to information and communication has come to replace a dissenting culture organized around the demonizing of abject hardware structures. Much, though not all, of the sixties counterculture was formed around what I have elsewhere called the *technology of folklore*—an expressive congeries of preindustrialist, agrarianist, Orientalist, and anti-technological ideas, values, and social structures. By contrast, the cybernetic countercultures of the nineties are already being formed around the *folklore of technology*—mythical feats of survivalism and resistance in a data-rich world of virtual environments and posthuman bodies—which is where many of the SF- and technology-conscious youth cultures have been assembling in recent years.[17] Some would argue, however, that the ideas and values of the sixties counterculture were only truly fulfilled in groups like the People's Computer Company, which ran Community Memory in Berkeley; or the Homebrew Computer Club, which pioneered personal microcomputing.[18] So, too, the Yippies had seen the need to form YIPL, the Youth International Party Line, devoted to "anarcho-technological" projects, which put out a newsletter called *TAP* (alternately the *Technological American Party* and the *Technological Assistance Program*). In its depoliticized form, which eschewed the kind of destructive "dark-side" hacking advocated in an earlier incarnation, *TAP* was eventually the progenitor of *2600*. A significant turning point, for example, was *TAP*'s decision not to publish plans for the hydrogen bomb (the *Progressive* did so)—bombs would destroy the phone system, which the *TAP* "phone phreaks" had an enthusiastic interest in maintaining.

There is no doubt that the hacking scene today makes countercultural activity more difficult to recognize and therefore to define as politically significant. It was much easier in the 1960s to *identify* the salient features and symbolic power of a romantic preindustrialist cultural politics in an advanced technological society, especially when the destructive evidence of

America's super-technological invasion of Vietnam was being screened daily. However, in a society whose technopolitical infrastructure depends increasingly upon greater surveillance, and where foreign wars are seen through the lens of laser-guided smart bombs, cybernetic activism necessarily relies on a much more covert politics of identity. Access to closed digital systems requires discretion and dissimulation, the authentication of a signature or pseudonym, not the identification of a real surveillable person, so there exists a crucial operative gap between authentication and identification. (As security systems move toward authenticating access through biological signatures—the biometric recording and measurement of physical characteristics such as palm or retinal prints, or vein patterns on the backs of hands—the hackers' staple method of systems entry through purloined passwords will be further challenged.) By the same token, cybernetic identity is never exhausted; it can be recreated, reassigned, and reconstructed with any number of different names and under different user accounts. In fact, most hacks or technocrimes go unnoticed or unreported for fear of publicizing the vulnerability of corporate security systems, especially when the hacks are performed by disgruntled employees taking their vengeance on management. So, too, authoritative identification of any individual hacker, whenever it occurs, is often the result of accidental leads rather than systematic detection. For example, Captain Midnight, the video pirate who commandeered a satellite a few years ago to interrupt broadcast TV viewing, was traced only because a member of the public reported a suspicious conversation heard over a crossed telephone line.

Eschewing its core constituency among the white male preprofessional–managerial class, the hacker community may be expanding its parameters outward. Hacking, for example, has become a feature of young-adult novel genres for girls.[19] The elitist class profile of the hacker prodigy as that of an undersocialized college nerd has become democratized and customized in recent years; it is no longer exclusively associated with institutionally acquired college expertise, and increasingly it dresses streetwise. In a recent article that documents the spread of the computer underground from college whiz-kids to a broader youth subculture termed "cyberpunks," after the movement among SF novelists, the original hacker phone phreak Captain Crunch is described as lamenting the fact that the cyberculture is no longer an "elite" one, and that hacker-valid information is much easier to obtain these days.[20]

For the most part, however, the self-defined hacker underground, like many other proto-countercultural tendencies, has been restricted to a privileged social milieu, further magnetized by its members, understanding that they are the apprentice architects of a future dominated by knowledge, expertise, and "smartness," whether human or digital. Consequently, it is clear that the hacker cyberculture is not a dropout culture; its disaffiliation from a domestic parent culture is often manifest in activities that answer, directly or indirectly, to the legitimate needs of industrial R&D. For example, this hacker culture celebrates high productivity, maverick forms of creative work energy, and an obsessive identification with on-line endurance (and endorphin highs)—all qualities that are valorized by the entrepreneurial codes of silicon futurism. In a critique of the hacker-as-rebel myth, Dennis Hayes debunks the political romance woven around the teenage hacker:

> They are typically white, upper-middle-class adolescents who have taken over the home computer (bought, subsidized, or tolerated by parents in the hope of cultivating computer literacy). Few are politically motivated although many express contempt for the "bureaucracies" that hamper their electronic journeys. Nearly all demand unfettered access to intricate and intriguing computer networks. In this, teenage hackers resemble an alienated shopping culture deprived of purchasing opportunities more than a terrorist network.[21]

While welcoming the sobriety of Hayes's critique, I am less willing to accept its assumptions about the political implications of hacker activities. Studies of youth subcultures (including those of a privileged middle-class formation) have taught us that the political meaning of certain forms of cultural "resistance" is notoriously difficult to read. These meanings are either highly coded or expressed indirectly through media—private peer languages, customized consumer styles, unorthodox leisure patterns, categories of insider knowledge and behavior—that have no fixed or inherent political significance. If cultural studies of this sort have proved anything, it is that the often symbolic, not wholly articulate, expressivity of a youth culture can seldom be translated directly into an articulate political philosophy. The significance of these cultures lies in their embryonic or *protopolitical* languages and technologies of opposition to dominant or parent systems of rules. If hackers lack a "cause," then they are certainly not the first youth culture to be characterized in this dismissive way: the left in particular has suffered from the lack of a cultural politics capable of recognizing the power of cultural expressions that do not wear a mature political commitment on their sleeves.

The escalation of activism in the professions in the last two decades has shown that it is a mistake simply to condemn the hacker impulse for its class constituency. To cede the "ability to know" on the grounds that elite groups will enjoy unjustly privileged access to technocratic knowledge is to cede too much of the future. Is it of no political significance at all that hackers' primary fantasies often involve the official computer systems of the police, armed forces, and defense and intelligence agencies? And that the rationale for their fantasies is unfailingly presented as a defense of civil liberties against the threat of centralized intelligence and military activities? Or is all of this merely a symptom of an apprentice elite's fledgling will to masculine power? The activities of the Chinese student elite in the pro-democracy movement have shown that unforeseen shifts in the political climate can produce startling new configurations of power and resistance. After Tiananmen Square, Party leaders found it imprudent to purge those high-tech engineer and computer cadres who alone could guarantee the future of any planned modernization program. On the other hand, the authorities rested uneasy knowing that each cadre (among the most activist groups in the student movement) is a potential hacker who can have the run of the communications house if and when he or she wants.

On the other hand, I do agree with Hayes's perception that the media have pursued their romance with the hacker at the cost of under-reporting the much greater challenge posed to corporate employers by their employees. Most high-tech "sabotage" takes place in the arena of conflicts between workers and management. In the ordinary, everyday life of office workers—mostly female—a widespread culture of unorganized sabotage accounts for infinitely more computer downtime and information loss every year than is caused by destructive "dark-side" hacking by celebrity cybernetic intruders. The sabotage, time theft, and strategic monkey-wrenching deployed by office workers in their engineered electromagnetic attacks on data storage and operating systems might range from the planting of time or logic bombs to the discreet use of electromagnetic Tesla coils or simple bodily friction: "Good old static electricity discharged from the fingertips probably accounts for close to half the disks and computers wiped out or down every year."[22] More skilled operators, intent on evening a score with management, often utilize sophisticated hacking techniques. In many cases, a coherent networking culture exists among console operators, where, among other things, tips about strategies for slowing down the pace of the work regime are circulated. While these threats from below are fully recognized in boardrooms, corporations dependent upon digital business machines are obviously unwilling to advertise how acutely vulnerable they actually are to this kind of sabotage. It is easy to imagine how organized computer activism could hold such companies to ransom. As Hayes points out, however, it is more difficult to mobilize any kind of labor movement organized upon such premises:

> Many are prepared to publicly oppose the countless dark legacies of the computer age: "electronic sweatshops," military technology, employee surveillance, genotoxic water, and zone depletion. Among those currently leading the opposition, however, it is apparently deemed "irresponsible" to recommend an active computerized resistance as a source of worker's power because it is perceived as a medium of employee crime and "terrorism."[23]

Processed World, the "magazine with a bad attitude," with which Hayes has been associated, is at the forefront of debating and circulating these questions among office workers, regularly tapping into the resentments borne out in on-the-job resistance.[24]

While only a small number of computer users would categorize themselves as "hackers," there are defensible reasons for extending the restricted definition of *hacking* down and across the caste hierarchy of systems analysts, designers, programmers, and operators to include all high-tech workers—no matter how inexpert—who can interrupt, upset, and redirect the smooth flow of structured communications that dictates their position in the social networks of exchange and determines the pace of their work schedules. To put it in these terms, however, is not to offer any universal definition of hacker agency. There are many social agents, for example, in job locations who are dependent upon the hope of technological *reskilling* and for whom sabotage or disruption of communicative rationality is of little use; for such people, definitions of hacking that are reconstructive, rather than deconstructive, are more appropriate. A good example is the crucial role of worker techno-literacy in the struggle of labor against automation and deskilling. When worker education classes in computer programming were discontinued by management at the Ford Rouge plant in Dearborn, Michigan, United Auto Workers members began to publish a newsletter called the *Amateur Computerist* to fill the gap.[25] Among the columnists and correspondents in the magazine have been veterans of the Flint sit-down strikes, who see a clear historical continuity between labor organization in the 1930s and automation and deskilling today. Workers' computer literacy is seen as essential, not only to the demystification of the computer and the reskilling of workers, but also to labor's capacity to intervene in decisions about new technologies that might result in shorter hours and thus in "work efficiency" rather than worker efficiency.

The three social locations I have mentioned above all express different class relations to technology: the location of an apprentice technical elite, conventionally associated with the term *hacking*; the location of the high-tech office worker, involved in "sabotage"; and the location of the shopfloor worker, whose future depends on technological reskilling. All therefore exhibit different ways of *claiming back* time dictated and appropriated by technological processes, and of establishing some form of independent control over the work relation so determined by the new technologies. All, then, fall under a broad understanding of the politics involved in any extended description of hacker activities.

THE CULTURE AND TECHNOLOGY QUESTION

Faced with these proliferating practices in the workplace, on the teenage cult fringe, and increasingly in mainstream entertainment, where over the last five years the cyberpunk sensibility in popular fiction, film, and television has caught the romance of the outlaw technology of human/machine interfaces, we are obliged, I think, to ask old questions about the new silicon order that the evangelists of information technology have been deliriously proclaiming for more than twenty years. The postindustrialists' picture of a world of freedom and abundance projects a bright millenarian future devoid of work drudgery and ecological degradation. This sunny social order, cybernetically wired up, is presented as an advanced evolutionary phase of society in accord with Enlightenment ideals of progress and rationality. By contrast, critics of this idealism see only a frightening advance in the technologies of social control—whose owners and sponsors are efficiently shaping a society, as Kevin Robins and Frank Webster put it, of "slaves without Athens" that is exactly the inverse of the "Athens without slaves" promised by the silicon positivists.[26] To counter the postindustrialists' millenarian picture of a postscarcity harmony in which citizens enjoy decentralized access to free-flowing information, it is necessary, then, to emphasize how and where actually existing cybernetic capitalism presents a gross caricature of such a postscarcity society.

One of the stories told by the critical left about new cultural technologies is that of monolithic, panoptical social control, effortlessly achieved through a smooth, endlessly interlocking system of surveillance networks. In this narrative, information technology is seen as the most despotic mode of domination yet, generating not just a revolution in capitalist production but also a revolution in living—"social Taylorism"— touching all cultural and social spheres in the home and the workplace.[27] Through gathering of information about transactions, consumer preferences, and creditworthiness, a harvest of information about any individual's whereabouts and movements, tastes, desires, contacts, friends, associates, and patterns of work and recreation becomes available in dossiers sold on the tradable information market, or is endlessly convertible into other forms of intelligence through computer-matching. Advanced pattern recognition technologies facilitate the process of surveillance, while data encryption protects it from public accountability.[28]

While the debate about privacy has triggered public consciousness about these excesses, the liberal discourse about ethics and damage control in which that debate has been conducted falls short of the more comprehensive analysis of social control and social management offered by left political economists who see information, increasingly, as the major site of capital accumulation in the world economy. What happens in the process by which information, gathered up by data-scavenging in the transactional sphere, is systematically converted into intelligence? A surplus value is created for use elsewhere. This surplus information value is more than is needed for public surveillance; it is often information, or intelligence, culled from consumer polling or statistical analysis of transactional behavior, that has no immediate use in the process of routine public surveillance. This surplus bureaucratic capital is used to forecast social futures, and consequently is applied to the task of managing in advance the future behavior of mass populations. As I will discuss in a later chapter, this surplus intelligence becomes the basis of a whole new industry of futures research that relies upon computer technology to simulate and forecast the shape, activity, and behavior of complex social systems. The result is a system of social management that far transcends the questions about surveillance that have been at the discursive center of the privacy debate.[29]

To challenge further the idealists' vision of postindustrial light and magic, we need only look inside the semiconductor workplace itself, home to the most toxic chemicals known to man (and woman, especially since women of color often make up the majority of the microelectronics labor force), where worker illness is measured not in quantities of blood spilled on the shopfloor but in the less visible forms of chromosome damage, miscarriages, premature deliveries, and severe birth defects. Semiconductor workers exhibit an occupational illness rate that, by the late 1970s, was already three times higher than that of manufacturing workers, at least until the federal rules for recognizing and defining injury levels were changed under the Reagan administration. Protection gear is designed to protect the product and the clean room from the workers, not vice versa. Recently, immunological health problems have begun to appear that can only be described as a kind of chemically induced AIDS, rendering the T-cells dysfunctional rather than depleting them like virally induced AIDS.[30] In corporate offices, where extraordinarily high stress patterns and illness rates are reported among VDT operators, the use of keystroke software to monitor and pace office workers has become a routine part of job performance evaluation programs. Some 70 per cent of corporations use electronic surveillance or other forms of quantitative monitoring of their workers. Every bodily movement, especially trips to the toilet, can be checked and measured. Federal deregulation has meant that the limits of employee workspace have in some government offices shrunk below that required by law for a two-hundred-pound laboratory pig.[31] Critics of the labor process seem to have sound reasons to believe that rationalization and quantification are at last entering their most primitive phase.

What I have been describing are some of the features of that critical left position—sometimes referred to as the "paranoid" position—on information technology which imagines or constructs a totalizing, monolithic picture of systematic domination. While this story is often characterized as conspiracy theory, its targets—technorationality, bureaucratic capitalism—are usually too abstract to fit the picture of a social order planned and shaped by a small, conspiring group of centralized power elites.

Although I believe that this story, when told inside and outside the classroom, for example, is an indispensable form of "consciousness-raising," it is not always and everywhere the best story to tell. While I am not comfortable with the "paranoid" labeling, I would argue that such narratives do little to discourage paranoia. The critical habit of finding unrelieved domination everywhere has certain consequences, one of which is to create a siege mentality, reinforcing the inertia, helplessness, and despair that such critiques set out to oppose in the first place. The result is a politics that can speak only from a victim's position. And when knowledge about surveillance is presented as systematic and infallible, self-censoring is sure to follow. In the psychosocial climate of fear and phobia aroused by the virus scare, there is a responsibility not to be alarmist or scared—especially when, as I have argued, such moments are profitably seized upon by the sponsors of control technology. In short, the picture of a seamlessly panoptical network of surveillance may be the result of a rather undemocratic, not to mention unsocialist, way of thinking, predicated upon the definition of people solely as victims. It echoes the old sociological models of mass society and mass culture, which cast the majority of society as passive and lobotomized in the face of modernization's cultural patterns. To emphasize, as Robins and Webster and others have done, the power of the new technologies to transform despotically the "rhythm, texture, and experience" of everyday life, and meet with no resistance in doing so, is not only to cleave, finally, to an epistemology of technological determinism, but also to dismiss the capacity of people to make their own use of new technologies, and to view technology as a contested site.[32]

The seamless "interlocking" of public and private information and intelligence networks is not as smooth and even as the critical school of hard domination would suggest. Compulsive gathering of information is no *guarantee* that any interpretive sense will be made of the files or dossiers. In any case, the centralized, "smart" supervision of an information gathering system would require, as Hans Magnus Enzensberger once argued, "a monitor that was bigger than the system itself"; "a linked series of communications . . . to the degree that it exceeds a certain critical size, can no longer be centrally controlled but only dealt with statistically."[33] Some would argue that the increasingly covert nature of surveillance indicates that the "campaign" for social control is not going well, and one of the most pervasive popular arguments against the panoptical intentions of technology's masters is that their systems do not work very well. Every successful hack or computer crime in some way reinforces the popular perception that information systems are not infallible. And the announcements of military–industrial spokespersons that the fully automated battlefield is on its way run up against an accumulated stock of popular skepticism about the operative capacity of weapons systems. These misgivings are born of decades of distrust for the plans and intentions of the military–industrial complex, and were quite evident in the widespread cynicism about the Strategic Defense Initiative. Just to take one example of unreliability, the military communications system worked so poorly and so farcically during the US invasion of Grenada that commanders had to call each other on payphones: ever since, the command-and-control code of Arpanet technocrats has been C^5—Command, Control, Communication, Computers, and Confusion.[34] The Gulf War saw the most concerted effort on the part of the US military–industrial–media complex to suppress evidence of such technical dysfunctions, which alone accounted, in the buildup to the war and its opening weeks, for much higher US casualty figures than those sustained in actual combat. The absence of the ineffective B-1 bomber, the most sophisticated weapons system of the 1980s, went largely unnoticed. As weeks went by, the crowing of the military and the public media about US supertechnology sounded more and more hollow. The Pentagon's vaunted information system proved no more—and often less—resourceful than the mental agility of its operators and analysts.

I am not suggesting that alternatives can be forged simply by encouraging disbelief in the infallibility of existing technologies. But technoskepticism, while not a *sufficient* condition for social change, is nonetheless a *necessary* condition. Stocks of popular technoskepticism are crucial to the task of eroding the legitimacy of those cultural values that prepare the way for new technological developments: values and principles such as the inevitability of material progress, the "emancipatory" domination of nature, the innovative autonomy of machines, the efficiency codes of pragmatism, and the linear juggernaut of liberal Enlightenment rationality—all increasingly under close critical scrutiny as a wave of ecological consciousness sweeps through the electorates of the West. Technologies do not shape or determine such values, which already preexist the technologies; the fact that they have become deeply embodied in the structure of popular needs and desires provides the green light for accepting certain kinds of technology. In fact, the principal rationale for introducing new technologies is that they answer to already existing intentions and demands that may be perceived as "subjective" but are never actually within the control of any single set of conspiring individuals. As Marike Finlay has argued, just as technology is possible only in given discursive situations (one of which is the desire of people to have it for reasons of empowerment) so capitalism is merely the site, and not the source, of the power that is often autonomously attributed to the owners and sponsors of technology.[35]

No frame of technological inevitability has not already interacted with popular needs and desires; no introduction of new machineries of control has not already been negotiated to some degree in the arena of popular consent. Thus the power to design architecture that incorporates different values must arise from the popular perception that existing technologies are not the only ones; nor are they the best when it comes to individual and collective empowerment. It was this perception—formed around the distrust of big, impersonal, "closed" hardware systems, and the desire for small, decentralized, interactive machines to facilitate interpersonal communication—that "built" the PC out of hacking expertise in the early seventies. These desires and distrusts were as much the partial "intentions" behind the development of microcomputing technology as deskilling, worker monitoring, and information gathering are the intentions behind the corporate use of that technology today. The machinery of countersurveillance is now up and running. The explosive growth of public data networks, bulletin-board systems and alternative information and media links, and the increasing cheapness of desktop publishing, satellite equipment, and international databases are as much the result of local political "intentions" as the fortified net of globally linked, restricted-access information systems is the intentional fantasy of those who seek to profit from centralized control. The picture that emerges from this mapping of intentions is not an inevitably technofascist one, but rather the uneven result of cultural struggles over values and meanings. These local advances are further assisted by the contradictions of capitalism itself, since market demand for ever cheaper and more resourceful technologies is putting video cameras and computing power into the hands of ordinary people who have traditionally experienced technology only as its object of surveillance, or, at best, its passive operator.

It is in the struggle over values and meanings that the work of cultural criticism takes on its special significance as a full participant in the debate about technology; a debate in which it is already fully implicated, if only because the culture and education industries are rapidly becoming integrated within the vast information service conglomerates. The media we study, the media we publish in, and the media we teach are increasingly part of the same tradable information sector. So too, our common intellectual discourse has been significantly affected by the recent debates about postmodernism (or culture in a postindustrial world) in which the euphoric, addictive thrill of the technological sublime has figured quite prominently. The high-speed technological fascination that is characteristic of the postmodern condition can be read, on the one hand, as a celebratory capitulation by intellectuals to the new information technocultures. On the other hand, this celebratory strain attests to the persuasive affect associated with the new cultural technologies, to their capacity (more powerful than that of their sponsors and promoters) to generate pleasure and gratification and to win the contest for intellectual as well as popular consent.

Another reason for the involvement of cultural critics in the technology debates has to do with our special critical knowledge of the way cultural meanings are produced—our knowledge about the politics of consumption and what is often called the politics of representation. This knowledge demonstrates that there are limits to the capacity of productive forces to shape and determine consciousness, insisting on the ideological or interpretive dimension of technology as a culture that can and must be used and consumed in a variety of ways not reducible to the intentions of any single source or producer; technology's meanings cannot simply be read off as evidence of faultless social reproduction. It is a knowledge, in short, that refuses to add to the "hard domination" picture of disenfranchised individuals watched over by some scheming panoptical intelligence. Far from being understood solely as the concrete hardware of sophisticated electronic objects, technology must be seen as a lived, interpretive practice for people in their everyday lives. To redefine the shape and form of that practice is to help create the need for new kinds of hardware and software.

One of this chapter's aims has been to describe and suggest a wider set of activities and social locations than is normally associated with the practice of hacking. If there is a challenge here for cultural critics, it might be the commitment to making our knowledge about technoculture into something like a hacker's knowledge, capable of penetrating existing systems of rationality that might otherwise be seen as infallible; a hacker's knowledge, capable of reskilling, and therefore of rewriting, the cultural programs and reprogramming the social values that make room for new technologies; a hacker's knowledge, capable also of generating new popular romances around the alternative uses of human ingenuity. If we are to take up that challenge we cannot afford to give up what technoliteracy we have acquired in deference to the vulgar faith that tells us it is always acquired in complicity and is thus contaminated by the toxin of instrumental rationality; or because we hear, often from the same quarters, that acquired technological competence simply glorifies the inhuman work ethic. Technoliteracy, for us, is the challenge to make a historical opportunity out of a historical necessity.

CHAPTER THREE

GETTING OUT OF THE GERNSBACK CONTINUUM

The title of this chapter refers to the cyberpunk writer William Gibson's first published story, "The Gernsback Continuum," whose narrator, a freelance photographer, has been hired to shoot 1930s futuristic North American architecture for the British nostalgia publishers' market. The streamlined design of factory buildings, gas stations, diners, and movie marquees—finned, flanged and fluted—recalls a future perfect that never was; a tomorrow's world planned and designed by technophiles faithful to the prewar ethic of progressive futurism. From the sleek rockets on the "spray-paint pulp utopias" of the Frank R. Paul covers of *Amazing Stories* magazines to the winged statues that guard the Hoover Dam, the story was of a promised future that would come into being only as a nightmare: the rockets were those that fell on London during the war; the streamlined cars and crystal superhighways gave the green light to postwar ecological atrocities committed by General Motors under the aegis of petroleum capitalism.[1]

In the course of his photo assignment, Gibson's narrator is haunted by the "semiotic ghost" of these outdated futures, chunks of "deep cultural imagery" from the "mass unconscious" that take on part-hallucinatory, part-material form as he travels around southern California—a flying-wing luxury liner, gleaming eighty-lane highways, and shark-fin roadsters; a city with ziggurats, zeppelin docks, giant neon spires; and a population of white, blond, "perfect" Americans, redolent of Nazi Youth propaganda. Advised by a friend to seek out "bad media" (porn movies, game shows, and soaps) to exorcize these ghosts, he plumps instead for the latest ecology-disaster literature on the newsstand, agreeing with the newsagent that things are bad, but assuring him that perfection would be—or would have been—much worse.

Gibson's elegant story is an economical commentary on the history of science fiction over the past half-century. Its appeal rests on a contrast between the tough, savvy realism of contemporary SF's fondness for technological dystopias and the wide-eyed idealism of the thirties pulp romance of utopian things to come. Through its sparing frame of reference to the generic history of science fiction, the story presents a state-of-the-art *explanation* of the current climate of anti-futuristic commentary, so much so, in fact, that it was selected as the lead story in *Mirrorshades* (1986), a definitive "cyberpunk anthology" edited by Bruce Sterling. Gibson's story, according to Sterling, "was a coolly accurate perception of the wrongheaded elements of the past—and a clarion call for a new SF esthetic of the Eighties."

Sterling's comment, clearly and unashamedly self-promoting, was part of an attempt to clear the ground and launch cyberpunk as a "movement" that marked a break with the past, and I shall have more to say about the cyberpunk moment in a later chapter. But the spirit of his dismissal of "the wrongheaded elements of the past" is widely shared by many within the SF community, embarrassed by the genre's origins in early pulp fiction. In the account of pulp fiction that follows, I will examine some of the reasons for this dismissal. As a counterbalance to the genre's sense of shame at its family origins, I will try to provide a more exhaustive historical explanation of the conditions of early SF, while arguing that many of its salient features, often considered to be "naive," were linked to central elements of progressive thought in the first three decades of this century.

Pop and camp nostalgia for the lofty ziggurats, teardrop automobiles, sleek ships of the airstream, and even the alien BEMs (bug-eyed monsters) with imperiled women in their clutches, are one thing; the cyberpunk critique of "wrongheadedness," whether in Gibson's elegant fiction or Sterling's flip criticism, is another. Each provides us with a stylized way of approaching SF's formative years, years that are usually described as "uncritical" in their outlook on technological progress. But neither perspective can give us much sense of the sociohistorical landscape of the thirties upon which these gleaming technofantasies were raised. To have some idea of the historical power of what Gibson calls the "Gernsback

Continuum," we need to know more, for example, about the entrepreneurial activities and scientific convictions of Hugo Gernsback himself, a man often termed the "father of science fiction" because he presided over its market specialization as a cultural genre (in Gernsback's view, SF was more a social than a literary movement). We need to know more about the hallowed place of engineers and scientists in public consciousness in the years of boom and crisis between the wars, the consolidation of industrial research science at the heart of corporate capitalism, and the redemptive role cast for technology in the drama of national recovery and growth. We also need to know about the traditions of progressive thought that stood behind the often radical technocratic philosophy of futuristic progress in the thirties. My description of the period of North American SF genre formation will show the crucial influence of national cults of science, engineering, and invention, as well as discuss technocracy's role in the social thought of the day. I will also consider the ways in which pulp SF escaped or resisted the recruitist role allotted to it not only by shaping figures like Gernsback, who devoted himself directly to enlisting his readers in the cause of "science," but also by subsequent critics of early SF—including those writers, like Gibson and Sterling, who have lamented its gung-ho celebration of technological innovation.

Because of their obligation to the conventions of historical narrative, literary historians of science fiction tend to favor linear accounts of its development as a cultural genre. Thumbnail sketches of its canonical literary history almost inevitably include the story of a formative naive period (Age of Gernsback and *Amazing Stories*), followed by the maturity of a Golden Age in the forties and fifties (Age of Campbell and *Astounding Stories*), succeeded by late modernist and proto-postmodernist challenges to the classical assumptions of the genre in the sixties, seventies, and eighties (New Wave, feminist-ecotopian, cyberpunk, etc.).[2] The advantages of this kind of narrative lie in its overall account of the formal history of the changing rules of generic conventions, understood and fully absorbed by historically conscious writers, readers, and fans within the SF community. Knowledge of these historical changes forms the "tradition" of a literary genre, inherited and used by its practitioners and cognoscenti. However, the linear conventions of this narrative tend to deflect our attention from the story—naive or not—that SF also tells about the place of science and technology in society at any one time. Consequently, a linear study is less useful to cultural historians who are interested in examining the role of SF in the material culture of a particular historical moment. It is especially difficult to write about the early years of a genre without falling prey to the knowledge that one is talking about "origins" that will then be "superseded" in the genre's later history. All historians confront this problem to some extent, but historians of a cultural genre are particularly constrained by knowledge of its subsequent development.

In the case of early SF, the conventional historical narrative is often overlaid by prejudices against the North American vulgarization of the high-minded and socially critical European SF tradition created by respectable intellectuals like H.G. Wells, Jules Verne, Aldous Huxley, Yevgeny Zamyatin, Fritz Lang, Olaf Stapledon, and Karel Capek. According to Brian Aldiss, the lowbrow American pulps were marketed as mere "propaganda for the wares of the inventor," where "screwdrivers substitute for vision" and where a standardized and "debased product" is churned out of "sweat-shops" of the mind.[3] For Aldiss, high European SF of the thirties grappled with the global zeitgeist— the rise of National Socialism and the fledgling specters of bureaucratic statism—while the lowly "bathetic" trade of the pulps was devoted solely to uncritical and unspeculative techno-jingoism, or to the crude, chauvinist Americanization of its largely immigrant readership.

Leaving aside the prejudicial bent of this description, one could still argue that Aldiss does the zeitgeist of the thirties a disservice by assuming that it flowed only through the refined synapses of European literary minds. The ideological backdrop to the pulps—the technocratic conflicts over the control and (mis)management of industrial production in a liberal democracy— harbored many of the ultimately decisive postwar solutions to the growing antagonisms between labor and capital. The postwar social contract that secured the long period of Fordist growth in the Western democracies proved to be a more resilient response to these antagonisms than the ideological cocktail of progressive modernization and reactionary folk nationalism that was the vehicle of European fascism, especially in the German version. In retrospect, technocratic Fordism and fascism were simply alternative solutions to the problems that capitalism faced in the thirties.

If the SF pulp authors of the twenties and thirties had little occasion or opportunity to directly address the highly charged role of technology in the social drama of their time, they nonetheless owed their literary *raison d'être* to the cults of science and technological invention that embellished a positivist religion (shared by left and right alike) in the years between the world wars. The pulp rhetoric of unstinting faith in the progressive virtues of science ranged across a wide political spectrum: the rhetoric was used by popular entrepreneurs like Gernsback, management reformers like F.W. Taylor, Progressive engineering professionals like Morris Cooke, business leaders and governing figures like Herbert Hoover and Theodore Roosevelt, non-Marxist technocrats like Thorstein Veblen and Lewis Mumford, and could even be found at the core of the organized revolutionary left, among those committed to the inevitability of "scientific socialism." To see this widely shared social fantasy as a naive example of blind faith in technological progress is not good enough; such judgments are part of our own naive response to history. Technocracy and Fordism were just as enthusiastically received by the European left (Antonio Gramsci says it all in "Americanism and Fordism"),[4] who saw these principles as applied modernist antidotes to fascist folk mysticism. To explain the significance of early SF, I will argue that we are obliged to substitute for the given wisdom about SF's "uncritical technophilia" a more historically nuanced account of its place in a context better described as *critical technocracy*.

So, too, any properly historical analysis of the early years of the genre ought to challenge the enduring assumption of cultural critics that pulp magazine production was an assembly-line culture designed to force-feed a passive population with a reactionary diet of escapist sensation. The market dynamics behind SF genre formation were indeed part and parcel of the pulp publishing revolution that created specialized audiences for detective fiction, spy fiction, horror fiction, Western fiction, romance fiction, and others in the interwar years. But SF deserves attention as a special case, if only because of the extraordinary role played by its amateur fans. It is still the only genre where the literary activities of its fans—in fanzines, gatherings, and conferences—far outnumber the published output of the genre's professional writers. Almost from the beginning, this obsessional amateur subculture came to mediate the magazines' appeal to a popular audience at every level, questioning, challenging, and contesting editorial power, determining the shape and success of certain magazines, and producing writers and editors from its ranks who, versed in the often politicized dialogue of fandom in the thirties, rapidly rose to transform the direction of the genre as a whole.

HUGO'S HARD LINE

To explicate fully what is meant by the "Gernsback Continuum," we need to look first at Gernsback's own role in publishing and in "fathering" the genre. Pulp publishing, said to have begun with Frank Munsey's all-story *Argosy* magazine in 1896, was a revolution in industrial mass production that capitalized on North America's vastly increased immigrant population, its decline in illiteracy rates, its rural postal delivery, and its increased urbanization. When pulp publishing was at its height in the late thirties, over 200 pulps with over 25 million readers were distributed across a wide social spectrum. Popular Publications, for instance, claims to have had both Al Capone and Harry Truman as subscribers.[5] Isaac Asimov recalls how his immigrant father, a candy-store owner, despised as trash the pulps he sold, but often read them to improve his English.[6]

In the pulp star system, it was usually the magazines, or their formulae, and seldom the writers themselves (however highly paid), who were the major actors. For the SF cognoscenti, the magazine editors played the starring role; consequently, it is Gernsback and John Campbell, editor of *Astounding Stories* from 1937 to 1971, who are commemorated as the "fathers" of science fiction, not Verne, Wells, Capek, or even E.E. "Doc" Smith, creator of the "space opera." It is clear, moreover, that Gernsback's "invention" of the genre, with the appearance of his pulp magazine, *Amazing Stories*, in 1926, is markedly different from, say, Munsey's fathering of the all-story pulp in the 1890s with *Argosy* and *All-Story Magazine*, or Street & Smith's creation of the specialized genre of crime-and-detection fiction with the publication of *Detective Story Magazine* in 1915. This is not to say that the private detective story, with its origins in Allan Pinkerton's dime novel memoir-adventures of the West, was any less an original "American" genre than the gadget-fetishizing Gernsbackian story, with its roots in the "invention hero" dime novels like the *Frank*

Reade Weekly Magazine, the ham-radio culture, and the popular boy-inventor culture of the turn of the century.[7] But unlike Munsey, an ex-telegraph operator turned publishing entrepreneur, Gernsback's own life story was exemplary of the starring role cast for the freelance inventor in the popular formulae he established as the generic "hard science" core of SF. In an age of collective corporate management, Gernsback's hands-on ownership and editorial role in the "invention" and early production of SF was a throwback to the myths of the individualist inventor–entrepreneur à la Edison of an earlier era. It is no coincidence that T. O'Conor Sloane, Gernsback's first managing editor at *Amazing Stories*, and an inventor and scientist in his own right, happened to be Edison's son-in-law as well.[8] Gernsback's rugged individualism went against the grain of the increasingly Taylorized culture industry, just as the inventor wizards who starred in the pulp SF stories had become anachronisms in the corporate research world of Bell Labs in the thirties.

As a promoter of the ham-radio craze in the first two decades of the century, Gernsback had already been at the center of another amateur culture long before he "created" the amateur fandom at the heart of science fiction culture. An immigrant from Luxembourg in 1904, and a sometime boy-inventor of layer batteries, his Electro Importing Company brought in specialized electrical equipment from Europe, but was more famous as the first mail-order radio house. He designed and sold the Telimco Wireless, the first home radio set (at a price so low his store was raided by the police) and the first walkie-talkie. He also began to publish a host of amateur science magazines. These, along with the Wireless Association of America founded by Gernsback in 1909, became the eyes and ears of the flourishing amateur radio movement. Relying on popular archetypes of the middle-class boy-inventor cult, which had inherited the romance of strenuous intellect from the likes of Frank Merriwell and Tom Swift, this fiercely networking community of boy ham operators played a significant role in contesting corporate and military attempts to establish monopoly control over the ether in the years before and after the First World War.[9]

By the time Gernsback's unwavering efforts to promote popular science came to include the first magazine devoted to science fiction in 1926, it was perhaps only with feigned surprise that, in the third issue of the magazine, he reacted to "the tremendous amount of mail we receive from—shall we call them 'Scientifiction Fans?' " In the prespecialized days of all-story magazines, it had already been established that there was a large readership for the kind of "gosh-wow!" stories that Gernsback printed (*Amazing* would soon have a readership of 100,000). As a result, his novel attempt to recruit for the "cause" of popular science through pulp literature generated a cult-like following of boys willing to be enlisted in the service of the new cause. In fact, it was the crusading zeal with which Gernsback's SF publications, from *Amazing* onwards, promoted scientific education and recruitment that served to demarcate SF from the whole field of fantasy stories out of which the SF story grew and that continued to flourish in competitor pulps like *Weird Tales* all through the twenties and thirties.

If the function of Gernsback's "scientifiction" was to consolidate popular education about science and technical knowledge, then the literary tendency to encourage formal "invention," especially of the sort that led to flights of fictional fancy, had to be subject to a restrictive principle. Indeed, the tension between technical and literary invention stretches across the whole history of science fiction: Verne once said famously of Wells's stories that they " 'do not repose on very scientific bases. . . . I make use of physics. He invents.' "[10] All popular genre fiction tends to have its own rules that *guard* against overinventiveness, but in the case of SF, the distinction between "using science" and "inventing" lay at the very heart of genre formation for other than formalistic reasons. To make his rules stick, to demarcate his genre of stories from those of his competitors, Gernsback created a whole supervisory apparatus to guarantee readers the "technical plausibility" and adherence to scientific fact of the prophetic stories he published under such slogans as "Extravagant Fiction Today—Cold Fact Tomorrow" (*Amazing Stories*); "Prophetic Fiction Is the Mother of Scientific Fact" (*Science Wonder Stories*); and "Adventures of Future Science" (*Wonder Stories*). Gernsback's editorial line in these matters can be summed up in this statement from the first issue of *Science Wonder Quarterly* in 1929: "It is the policy of *Science Wonder Stories* to publish only such stories that have their basis in scientific laws as we know them, or in the logical deduction of new laws from what we know."[11]

How was this policy regulated and what did it mean in literary practice? Not only did Gernsback establish a panel of experts—all reputable professionals from universities, museums and institutes—to pass judgment on the accuracy of the science; he also encouraged his writers to elaborate on scientific details they employed in their stories and to comment on the impossibilities in each other's stories. He even offered his readers prize money for identifying scientific errors. While minor errors were forgiven, writers not up on science were quickly dropped. Factual, nonfiction articles about science and technology could be found in each issue of his magazines, and the editors' column, a regular feature, provided a forum for readers' queries about the world of science ("Science Questions and Answers," imitated by competitors like *Astounding Stories*' "Science Forum"). Gernsback's readers were addressed as a special breed: they were young men with high school educations who had a hobby life as amateur chemists, astronomers, radio novices and the like. They were likely to be reading Gernsback's companion publications like *Science and Invention* and *Radio News* for the purpose of self-improvement while other pulp aficionados perused more lurid sensation in such fare as *Ranch Romances*, *Ace-High*, *Spicy Mystery Stories*, or, heaven forbid, *Love Story* and *Miss 1930*.

Gernsback also fought long and hard, in a highly competitive arena, to permanently attach his agenda of technical plausibility and hard science to a generic term for the specialty field. This term, "science fiction," emerged only after a long contest waged between competing magazines like *Argosy*, *Weird Tales*, and *Astounding Stories* over terms like "pseudoscientific stories," "scientifiction," "weird-scientific stories," "off-trail stories," "fantascience," "super science." and others.[12] Of course, the contest over the definitive term had high commercial stakes in the subscription game, but it was also part of a search, in the pulp world, for a stable, legitimate standard around which the loyalty of fans and readers could be mobilized. Long after 1934, the year when *Astounding Stories*, Gernsback's chief competitor, was reorganized and began to dominate the field, Gernsback's rules of play concerning the centrality of the hard physical sciences continued to hold sway. Story lines diversified. Orlin Tremaine, then the editor of *Astounding*, introduced the more metaphysical "thought-variant" story. John Campbell, Tremaine's maverick successor, encouraged his famous stable of writers to try more speculative, psychohistorical, and even sociological treatments. In the fifties, Campbell allowed his writers to investigate the fields of psi, dianetics, and parapsychology. Even in Gernsback's heyday, a number of fanciful scientific "errors" (like hyperspace) were tolerated as *superscience* conventions in order to explain the interstellar plausibility of the generic "space opera," pioneered by E.E. Smith in *The Skylark of Space* (1928). For the most part, however, the gatekeeper-editors at the head of the field stuck to the positivist line as an issue of fundamental policy until well into the fifties, when dystopias and critiques of the religion of "progress" through science and technology eventually began to predominate in a field founded on the idea of progressive futurism, the dominant discourse of its day. (One of the chief exceptions was Ray Palmer—famous among fans, ironically, for winning Gernsbackian contests—who rose from fandom to be editor of the post-Gernsback *Amazing* in 1938, and who outmatched *Astounding*'s circulation figures after the war by encouraging the cult of irrationality and UFOism through Richard Shaver's stories about Lemuria.)

Gernsback often used the fandom he had helped create to regulate these hard science policies. For example, the second issue of *Amazing* carried a Gernsback editorial that cited a letter from one George Anderson of Fairmount, West Virginia, suggesting that the magazine should "print all scientific facts as related in the stories in italics. This will serve to more forcefully drive home the idea upon which you have established your magazine."[13] Other readers complained that the stories were too scientific, and that more attention ought to be paid to literary style. Although neither suggestion was really taken up, the typical Gernsbackian story of adventures through gadgetry always featured moments in which the genius–inventor took time to explain at length—often in isolated and stylistically undigested paragraphs—the science that he was employing to save the world.

Critics of Gernsbackianism have charged that Gernsback's devotion to the pragmatic, hardware-oriented tradition of invention was a formula for technological fiction only and had little to do with a properly scientific fiction that fully questioned the nature of the objective world. In the years before SF was established generically, the new quantum physics, for example, had been exploring the heady qualities of a newly implausible universe. At the core of the Gernsback formula, however, was a populist principle that science could be explained and understood by everyone, and that its name would not be associated with exclusive rhetorical idioms or with obfuscatory accounts of the object world by overaccredited experts.

For Gernsback, scientific language was a universal language of progress that ought to be accessible even to those without a college degree. Indeed, the straightforward prose of early SF clearly contrasts with the rich American argot of local dialects found in the Western and hard-boiled genres. Its undeveloped style is equally distinct from the luxurious hand of the likes of H.P. Lovecraft, whose overwritten fantasies crammed the pages of *Weird Tales*.

All the same, there were dialects to be heard in Gernsbackian fiction, specifically those of the alien species who were typically vilified in the most overtly racist ways. Since women—who featured variously in *Weird Tales* as vampires, witches, and high priestesses—rarely made an appearance in pulp SF (Frank Paul, Gernsback's illustrator, was a wizard at drawing men, machines, and monsters, but was said to be technically deficient when it came to depicting female bodies, especially the blonde Amazons who would feature heavily on pulp covers after the thirties) this "universal" language of science was, in practice, for white boys and men only. In particular, this language functioned as a mark of the heroes' superiority in coping with conditions in exotic localities like Mars or Venus, whose climates were simply displaced from the popular action-adventure regions of the arid West and tropical Africa respectively. So, too, the jargon of "positronic rays" and "electronic vibration adjusters" quickly became the mark of an insider language that readers and fans could learn to access and cultivate as the language of experts, eventually producing their own subcultural variant in the unique idiom of fanspeak.

Although the "universal" language of science and rationality popularized in pulp SF was tailored to a rather narrow, white-male constituency, it could still be construed as a populist refusal of the elitist vehicles of "literary" speech and "metaphysical" discourse that had traditionally dominated Western literate culture. Indeed, the belated recognition of SF as a literary genre by technophobic humanists was an effect of its perceived challenge to that tradition of humane discourse. The spare, economical language of technorationality, everywhere valorized in the twenties and thirties as the official language of the latest version of North American pragmatism, embodied the austerity measures also favored by much of high modernist culture. One thinks of the ascendancy of the maxim "form follows function" in art, design, and the modern movement in architecture; or the rhetorical economy of Ezra Pound's Imagism, in which excess was condemned as "wasteful." Many of the formal principles that lay behind the modernist movement can be seen as a literal translation of the efficiency techniques of Taylorism. The new technocratic principles of stark efficiency, tight economy, and hard precision had come to fill the role vacated by the eschewal of ornament and the rejection of conspicuous expenditure associated with the wasteful style of late-nineteenth-century high bourgeois culture. These utilitarian conventions were broadly welcomed, in culture and social thought, for their appeal to the principles of democratic modernization.

The history of the adoption and incorporation of utilitarian conventions into antidemocratic philosophies in the course of the 1930s is well known, especially in the case of European high modernism, many of whose adherents followed the road to fascism without having to deviate from their politics of style. In Germany, where a specialized mass genre of SF did not exist, Nazi ideology was nourished by the Aryan-mythological elements of much of the "Teutonic" fantasy fiction, by the purified spiritual histories espoused by various occult science groups, and even by the activities of the Society for Space-Flight, whose amateur experiments with rocketry were lionized in the American pulps throughout the thirties. The National Socialist fusion of precapitalist pastoralism, technological modernization and millenarian futurism were all fully present in the German SF culture of the time.[14]

Arguably, however, the best examples of Nordic–Aryan adventure fiction could be found in the North American pulps during these years. The historical continuity of these stories with the colonial romance of the Western served to reaffirm whatever codes of nationalistic destiny were at work in the formula. The colonial romances of the new high-tech adventure formula had to take place in foreign locations; they could not be pursued in the squalid urban conditions familiar to the modern, techno-intensive labor force. In this respect, North American SF was much more than a naive reflection of the cult of technology; it was also an embryonic response to the call for the colonization of space, where adventure, as always, was imperialism's accomplice. The language of colonization gave the sparse instrumentality of scientific boyspeak an American accent, just as the lingua franca of science among earthlings today is American English.

In the fledgling struggles over genre formation in the SF of the twenties and thirties, what we can see is a contest to establish a language that signified scientific rationality and to eliminate a language that privileged romance, fantasy, and literary invention—except, of course, where the romance was that of science and technology.[15] Popular SF was as actively committed to this crusade as the modernist movement was to purging the rhetorical vestiges of romanticism. Today, the opposition between science and rhetoric is no longer as clearly defined. While the original story of science is still told in opposition to the humanist tradition of rhetoric, in recent years critics have come to see science itself as just another form of rhetoric; one with particularly aggressive claims on objectivity. In literary practice, of course, Gernsback's purist devotion to the rhetoric of hard science was everywhere open to adulteration, as the genre could not stand still if it was to be true to the spirit of fictional, or even scientific, invention. A literary commodity that traded on its "amazing" and "astounding" qualities was even less likely than other forms to be able to toe the hard line against rhetorical excess.

The now famous pulp cover illustrations of Frank R. Paul and others offer an alternative graphic example of the rhetorical practice found in these early stories. Paul's designs were intimate with the Streamline Moderne style, the sexy, populist American answer to the European modernist aesthetic that embodied the new functionalist philosophy.[16] In contrast to the ascetic feel of Bauhaus modernism, streamlining took a more playful, stylistically expressive approach to the cult of progressive paring. Lifting the parabolic curve from aerodynamic design, streamline's ability to lend dynamic movement to even the most motionless objects (Raymond Loewy's pencil-sharpener was the most controversial example) differed from the more severe, stationary aesthetic of modernist geometrical design. Excoriated by the priests of functionalism for its decorative, "superficial" styling, streamlining was more than just a populist version of the "less-is-more" aesthetic. In the thirties, it became the "look" of the future, if only because streamlined objects always looked as if they were going somewhere; quietly, rapidly, smoothly, and hygienically. The industrial logic of obsolescence, which streamline design helped to introduce, promised, through yearly stylistic changes and advances, that the object world was steadily and surely moving toward a state of technical perfection.

Despite editorial reminders about the genre's commitment to hard facts and equations, pulp SF shared the same kind of populist aesthetic as streamline, and the cover art advertised the same futurist promise of a genre that was going somewhere fast. SF would very soon become part of the social knowledge of public culture. If that were not already apparent in the public response to the Mercury Theater's "War of the Worlds" radio broadcast in 1938 (bolstered, no doubt, by public paranoia about the technically advanced airborne exploits of the Wehrmacht), then it became all too clear, as I will describe later, in the worshipful respect with which visitors were encouraged to visit the streamlined World of Tomorrow at the 1939 New York World's Fair—as if to drive home the message that they weren't in the "Kansas" of *Amazing Stories* any more.

SCIENCE FICTION ACTION

It has been argued, with good reason, that Gernsback's policy of technical plausibility held back some of the "more imaginative and socially aware writers in the magazine field for more than a decade."[17] Many of these writers could nonetheless be found active in the ranks of SF fandom in the thirties, where the romance of the boy fiction-inventor had begun to run much deeper than the romance of the boy hardware-inventor. Of course, this tendency ran counter to the Gernsbackian recruitment philosophy, reaffirmed in his creation of the Science Correspondence Club for amateur scientists in 1930 (later, the International Scientific Association), which was devoted to "the furtherance of science and its dissemination among the laymen of the world and the final betterment of humanity." By the time Gernsback formed the first national fan association, the Science Fiction League, in 1934, primarily to boost subscriptions during the Depression, the tenor of amateur correspondence and networking was firmly tied to the goal of active critical participation in the whole adventure of the SF genre. While some of the local chapters of the Science Fiction League (SFL) toed the hard science line—the acid test seemed to be an active interest in launching rockets, a feat achieved by not a few clubs—others engaged in active criticism of the prozine publishers, editors, and formats through the medium of their own fanzines. So too, the splits, dissensions, expulsions, purges, and heated conflicts that characterized the activity of the numerous groups in the course of the first decade of fandom mimicked the sectarian conflicts being waged in the political groupings of the organized left throughout the thirties.

In fact, the Futurians, the most famous group of fans, included some Young Communist Leaguers who, not unlike Gernsback, approached fandom as a recruitment opportunity for their own socialist cause. The Futurians eventually came to include such fan-writers as Donald Wollheim, John Michel, Frederik Pohl, Cyril Kornbluth, Isaac Asimov, Jack Gillespie, Damon Knight, Robert Lowndes, Dirk Wylie, Richard Wilson, James Blish, and the first women in fandom, including Judith Merril, Doris Baumgart, and Rosalind Cohen. One of the origins of this grouping lay in a visit by Wollheim to local New York SFL chapters to verify if other members shared his own experience of contributing to Gernsback's magazines but never being paid for his work. Out of this gesture towards labor unity grew small pre-Futurian groups like the Committee for the Political Advancement of Science Fiction, which published the *Science Fiction Advance* fanzine. They also sponsored John Michel's famous political interventions at the first national fan conventions—the "Mutation or Death" speech at Philadelphia in 1937, modeled with wit on a Tremaine "thought-variant" story—and the less lively address, suppressed but circulated at Newark the following year, entitled: "The Position of Science Correlative to Science Fiction and the Present and Developing International Economic, Political, Social and Cultural Crisis."

Michelism, as it came to be known, was "the theory of science-fiction action," tied to the socialist ideal of the science of social "progress." It therefore embodied a much broader conception of science than that encompassed by Gernsback's philosophy. In a 1938 issue of the British fanzine *Terrae Novae*, Wollheim bluntly expounded the base principle of Michelism, demonstrating, perhaps, what he and Michel nonetheless shared with Gernsback in seeing SF as a medium for recruitment.

> MICHELISM is the belief that SF followers should actively work for the realization of the scientific socialist world-state as the only genuine justification for their activities and existence.
> MICHELISM believes that science-fiction is a force; a force acting through the medium of speculative and prophetic fiction on the minds of idealist youth;

> that logical science-fiction inevitably points to the necessity for socialism, the advance of science, and the world-state; and that these aims, created by science-fictional idealizing, can best be reached through adherence to the program of the Communist International.[18]

If statements of Michelist principle, like this one, could be leaden-footed, the Futurians were just as capable of appealing on SF grounds, basing their recruitment tactics on arguments drawn from well-known SF works in which fascist conditions were an element of future worlds—there was always H.G. Wells to invoke!

The heated debate about Michelism absorbed the world of fandom for some time, and while the Futurians lost the contest for supremacy in the national fan organizations, and the much sought after internationalist control over WorldCon, their injection of social consciousness into the fandom world had an enduring effect at a time when the pulp stories were beginning to address the future of authoritarian social orders. Graduating to the ranks of professional editors and writers at the end of the decade, they eventually formed something of a counterculture operating against the established power of the field's publishers and editors. Having lived communally for a number of years, their collaborative writing habits bore fruit—especially in the case of Kornbluth and Pohl, whose novels like *The Space Merchants* (1953), became classics of socially critical SF. As a mutual self-help community, they also followed the practice of giving their stories for free to the professional magazines now edited by comrades like Pohl and Wollheim, a practice that establishment publishers of the day openly redbaited.[19]

In 1940, during the period of the Soviet–Nazi pact, a number of Futurians took a belated interest in the Technocracy Study Course, which offered the thought of Howard Scott, the leading figure in the Technocracy movement in the thirties. As Lowndes rather obscurely put it: "We were Stalinists disguised as Technocrats. We went into it for the purpose of a cover. We became very subdued and wound up as progressive liberals."[20] Whatever the motives of the Futurians and whatever the outcome of their interest—they dropped Scott the following year when Hitler invaded the Soviet Union—they had made official contact with the scenic world of Technocracy Inc., one of the most quirky, but significant, non-Marxist movements to emerge from the anti-capitalist social philosophy that had fashioned a religion out of the progressive uses of science and technology in the previous three decades.

ENGINEERS' DREAMS

In 1940, Technocracy was hardly a new idea to the science fiction world. Some of the important West Coast fan clubs had been openly Technocrat. Gernsback himself had briefly edited a magazine called *Technocracy Review* during the heyday of the movement in 1933.[21] His editorials in *Wonder Stories* at the time praised the foresight of earlier SF in developing many of the ideas that lay behind the Technocracy movement, and he commissioned Nathan Schachner to write a series of articles about the movement in the magazine. The heroes who populated the prehistory of the Technocracy movement—the romantic engineer–adventurers of popular film and fiction like *Soldiers of Fortune*, *The Trail of the Lonesome Pine*, *Fighting Engineers*, *The Winning of Barbara Worth*[22]—were figures who could have been found in almost any SF pulp story over the previous fifteen years. They were figures not unlike the modern Indiana Jones, wearing broad-brimmed hats, heavy, high-laced boots, riding breeches, big leather jackets, red bandanas, and well equipped with rolls of blueprints and slide rules.

Howard Scott, the architect of Technocracy, and a somewhat bogus engineer himself, had dressed like this all through the twenties, seducing intellectuals with his rough-hewn image and his prophetic Greenwich Village coffee-house talk about increasing the efficiency of national production. By the end of the thirties, however, he had long since traded the leather jacket for a uniform gray flannel suit of the sort that came to be associated with the "organization man" of the postwar period, and whose drab features are closer to the prosaic connotations of the term "technocrat" today. In Scott's case, however, it was a regulation dress (with a blue necktie, no doubt to connote a relation, however distant, to labor), tailored to specifications and worn nationwide by the Technocrats, who had developed their own weird proto-fascist trappings (while being strictly anti-fascist); they sported armbands and lapel pins bearing the organization symbol, the Monad, and boasted their own technical organizations like the Technocrat motorcycle and automobile corps, along with a youth group called the Farads. Scott took to recognizing a paramilitary salute in the course of his travels to address followers in the regional groups who, at their height, may have numbered as many as 250,000: most of them native-born, déclassé, white Anglo professionals (a large percentage female) who were left populists, or socialists alienated from the organized left parties.[23]

Over two decades, then, the dominant image associated with the technocratic idea had shifted from that of self-sacrificing, rugged individualism to that of an organized, millenarian movement with showy proto-fascist trappings. For a set of ideas whose proponents saw themselves as having gone "beyond politics" in their allegiance to scientific thought, this shift was nothing if not responsive to the political zeitgeist. The story of the technocratic idea was, in many respects, more representative of domestic US culture in the twenties and thirties than the romance of American Communism, although it is the less well known of the two. Arguably, the appeal of Technocracy to native populism and to philosophical pragmatism (jokes about the "dictatorship of the engineers" notwithstanding) held more sway over popular anti-business consciousness than the pre-Popular Front CPUSA image of a "Soviet America," so captivating to Europeanized American intellectuals of the time. Compared to the crucial, though involuntary, scapegoat role played by American Communism at the heart of official US ideology for almost three decades, the influence on national politics of Technocracy's only organized movement, Scott's Technocracy Inc., was almost negligible.

Nonetheless, the short-lived moment of Technocracy—hyped as a "solution" to the Depression in a storm of media attention in 1933–34, then just as quickly vilified by business economists for its "pseudoscientific" blueprint for abolishing the "price system" of capitalism—was one in which many of the right questions were being asked about a technically advanced society with the capacity and resources to reduce the burden of human labor and the squandering of resources to a fraction of their current levels. Under the aegis of slogans like "governance by science, social control through the power of technique," the Technocrats were directly addressing issues like automation and technological unemployment, the rationalized use of expertise and management over the rule of capital, industrial democracy, production for use and not for profit, and the non-utopian horizon of a postscarcity culture. The solution it offered to capitalist mismanagement of industrial production was quite eccentric and impractical—a complete replacement of the economy's dependence on commodity value with a system based on available energy resources and technologies, wherein the unit of value would be energy (the erg, joule, or calorie) and not monetary (the dollar).

Scott and a number of Columbia academics had formed the Committee on Technocracy in 1932 to produce an Energy Survey of North America, which included an analysis of the energy expenditure of 3000 products and a survey of the physical functions of the population relative to the thermal units produced and consumed. The research team concluded that national production was hopelessly inefficient, and that the price system was to blame: "if the total one billion installed horse power of the United States were operated to full capacity, its output would be equivalent to the human labor of over five times the present world population." They concluded that if all available energy resources could be converted into use-value, then capitalist problems of overproduction, underconsumption and unemployment would disappear. Advocates claimed that Technocracy's quantitative social philosophy, with its scientific technique for arriving at all decisions, was the only program honestly devoted to the postscarcity future that was the hollow promise of consumer capitalism. All other programs were antiquated, especially Marxism, which, in Scott's view, was a mere philosophy, not a science, and a "scarcity philosophy," at that; he called it "an intellectual expression of dementia praecox."[24]

Scott's movement lived and died by the sword of science, when its statistics were successfully challenged. More important, however, his own fate was that of middlebrow intellectuals who risk legitimizing their claims by appealing to accredited knowledge, but who lack legitimate accreditation themselves. As it happens, his rhetoric had all the obfuscatory trappings of high science and very little populist bite:

> Technocracy makes one basic postulate: that the phenomena involved in the functional operation of a social mechanism are metrical. It defines science as the "methodology of the determination of the most probable." Technocracy therefore assumes from its postulate that there already exist fundamental and arbitrary units which, in conjunction with derived units, can be extended to form a new and basic method for the quantitative analysis and determination

> of the next most probable state of any social mechanism. Technocracy further states that, as all organic and inorganic mechanisms involved in the operation of the social macrocosm are energy-consuming devices. . . . Technocracy accordingly establishes a new technique of social mensuration, that is to say, a process for determining the rates of growth of all energy-consuming devices within the limits of the next most probable energy state.[25]

One would have to look hard to find a better prototype from this period of the kind of rationalist technospeak that has increasingly come to dominate the bureaucracies of North American and transnational business, government, education, and military diplomacy. Today this language signifies impersonality and inhumanity, the technocrat's passionless cross to bear. In the twenties and thirties, it was the language of modernity and progress, cutting through the elitist value-system that championed the extravagant carelessness of the "one-eyed" captains of industry and granted "pecuniary distinction," in Veblen's terms, to the genteel rhetorical excesses of their beneficiaries, the "kept classes." All advocates of social action were obliged to emulate this language in some way, for it was the language of efficiency—even when it seemed long-winded and opaque, as Scott's does. The shared claim of the countercultures of the day—whether socialist, technocrat, populist or avant garde—was that they could speak this language more proficiently than the dominant culture. While sharing the dominant values of modernity and progress, these countercultures promised to be more creative, more productive, more efficient, more growth-oriented, and more humane than the fettered capitalist management of society was proving to be.

While elements of the Technocracy program were quickly absorbed and incorporated into corporate business culture, the obscure fate of the movement itself was a foregone conclusion in the political combat zone of the thirties. Overly satisfied, like some Marxists, by the scientific complacency of a doctrine that expected the price system imminently to collapse and fade away, the Technocrats consequently had no pragmatic theory about the "assumption of power." As technological determinists and subscribers to a belief in the rationality of expert decision-making, they had no real need for *politics* (as irrational as business), let alone a requirement for democratic politics. Neither was the scientism of Technocracy supposed to be "value-oriented," and so it rejected any utopianism built on moral or political value-systems. This tendency caused a split in the movement between Scott's supporters and members of the more humanist Continental Committee like Harold Loeb, author of the utopian novel *Life in a Technocracy* (1938). The deepest cut of all, for utopians like Loeb, was that nonfunctional activities—the ethical, cultural and intellectual life of the mind and body—were all eliminated under the Technate, Scott's futurist plan for reorganizing the "social mechanism" of the North American states.[26] The resulting social picture was the kind of proto-dystopian future order from which science fiction would draw its own elemental version of the conformist bureaucratic state in which citizens enjoy a torpid, chloroformed life of robotic efficiency and bloodless rationality.

In the postwar years that lay beyond Technocracy, many of its tendencies took material shape: the rise of the "organization man" in the management technostructure, the triumph of quantification as a dominant social philosophy, the announcement of the "end of ideology," and the Fordist compromise between capital and labor, which temporarily balanced the needs of capitalist technological growth against the demands of labor for a social wage. What was left out of the Fordist bargain, however, was any consideration of non-work, in many respects the radical founding initiative of the Technocracy movement.

Forged in the crisis years of the Depression, Technocracy had insisted that it was capitalism (notwithstanding the euphemism "price system") and not technology that was to blame for the massive unemployment, poverty, and starvation; and that it was a mistake to see the technological means of production as independent from the social relations of production. In this, as in many other theoretical convictions, it was faithful to the progressivism of the engineers' reform movement from which it derived. At its roots was the vanguardist role Thorstein Veblen cast for the engineer. Going back even further, there were links with the Midwestern agrarian populists who had excoriated absentee landlords for their distant mismanagement of *real* agricultural production.

Edwin Layton, historian of the earlier Progressive engineers' movement in the teens, rejects Technocracy as "a grotesque parody of the engineers' thought, rather than a legitimate expression of it."[27] But Scott and his associates rose to address the crisis conditions of the Depression in a way the professional engineers, who had long since made their uneasy peace with capital, were no longer in a position to do. For the most part, the latter were content to patronize the large-scale, giant power schemes of the New Deal, like the TVA and Rural Electrification, headed up or staffed by former Progressives like Morris Cooke. On the other hand, there is no doubt about Technocracy's debt to the doctrine of social responsibility that, in earlier decades, had promoted the professional class of engineers, as stewards of technology and social change, to the forefront of reformist and progressive thinking. Engineering had become a messianic vocation in the first three decades of the century. It was a time when technological progress was raised to the status of a self-evident truth, and when the cults of efficiency and waste conservation presided over everyday life. Even women in the home, as Cecelia Tichi has described, were addressed as the engineers or scientific managers of their households: toward the end of the teens, "Training the Home Engineer" and "Running the Home Like a Factory" were typical lead articles in popular magazines like *Women's Home Companion*.[28]

Beholden to the primacy of scientific principles, the engineer was called upon to be a functional architect of modernity along the lines of utility, not profit. Progressives, conservatives, and Taylorist advocates of scientific management all shared the rhetoric of conservation and efficiency but differed, of course, on the radical role of the engineer's commitment to turn a blind eye to the golden rule of profit. Unlike lawyers and doctors, most engineers were not self-employed, and so attempts to radicalize the profession in the teens were hard put to challenge the influence of their capitalist keepers. Nonetheless, popular consciousness about the wastefulness of financiers and the "lag, leak and friction" of industry eventually ran so high that the engineer Herbert Hoover ran a successful presidential election campaign imbued with the demand for national efficiency, a clear sign that the reformist spirit had entered business culture itself.

More radical was the idea, embodied in Veblen's plan for a "soviet of technicians," that engineers could form a visionary avant garde to sweep away the inefficient price system and usher in maximum production for use.[29] With a practical agenda in mind, a Technical Alliance of economists, engineers, accountants and physicists, which included both Veblen and Scott, along with Walter Rautenstrauch and Guido Marx, was formed at the New School in the early twenties. The report of their Committee on the Elimination of Waste presented a devastating critique of employer mismanagement, while a second report on the labor productivity of the twelve-hour day recommended an eight-hour day, again in the name of industrial efficiency.

So pervasive was the moral category of efficiency that Veblen's revolutionary "soviet of technicians" was received as a respectable proposition in many circles in 1920. More to the point, his engineer avant garde was much safer, from the perspective of business interests, than the revolutionary power of labor, to which the engineers were quite hostile (class conflict was too "wasteful" to be scientific). As scientific management centralized and concentrated power in the hands of hireling experts, it increasingly shifted attention away from demands for self-government or any form of democratic self-management. The status-seeking role of the professional bodies, fired by a revolution of rising expectations about the market value of their members' skills, would easily be reconciled with loyalty to business interests, while the superhuman features of engineer ideology (like those technological powers celebrated in the comic-book superheroes of the late thirties) would ultimately be used to mystify, rather than socialize or radicalize, the profession. Hoover, the engineer President, popularized the term "rugged individualism," at a time when the vestigial romance of the engineer still traded on that quality, but when most engineers had long since become lackluster corporate retainers.

In fact, the heavily mythologized story of rugged, reformist individuals' domestication by corporate rationality is everywhere underpinned by the history of the industrial incorporation of science itself. Veblen and his followers came to pose science and business as good and bad angels respectively: two conflicting demands upon the engineer's vocational conscience. By this time, however, science had long been the governing genius of industrial production; technological progress, as applied science, had become the primary rationale for capitalist growth. The crisis tendencies of capitalist overproduction were now held in check, of course, by the principles of scientific management. In other words, there was virtually no *de facto* opposition between business and science. By the twenties, as David Noble has argued, not only had scientific knowledge come to govern the process of production, but the goal of modern industrial use of technology had become one of transforming science into capital:

> From the start, modern technology was nothing more or less than the transformation of science into a means of capital accumulation, through the application of discoveries in physics and chemistry to the processes of commodity production. . . . Modern science-based industry [was] industrial enterprise in which ongoing scientific investigation and the systematic application of scientific knowledge to the process of commodity production [had] become routine parts of the operation.[30]

In harnessing the myth of scientific progress through technological innovation, control of science itself had become an industrial monopoly, confined to the new corporate research laboratories, or to university locations where research worked hand in hand with corporate interests, and where technical education was shaped by industrial needs. Industrial corporations either controlled or purchased most invention patents, hoarding them in order to suppress competition, while independent inventors lacked the big capital that was increasingly required for research and development of their alternative technologies. As a result, the explosive age of industrial invention lay in the past, viewed as much too volatile in its effects for the scientifically regulated processes of production favored by monopoly capitalism. The new emphasis on control, precision, uniformity, predictability and standardization meant the extinction of the entrepreneur–inventor, whose heyday had included the likes of George Westinghouse, Francis Richards, Edward Weston, Alexander Graham Bell, Thomas Edison, Lee De Forest, Elihu Thomson, Charles Brush, Charles Hall, Edward Dean Acheson, and Elmer Sperry. While scientists became increasingly "proletarianized," the new Fordist compromise meant that monopoly cartels, often in agreement with labor unions, would seek new uses for old technology rather than risking the potential instability offered by new labor-saving, and ecologically sound, technologies.

Given these developments, it is easy to see how the cultural myth of the individual inventor (a mythic figure, even in the time of Edison, who controlled the markets created by his inventions and who pioneered the incorporation of science as shrewdly as anyone), like that of the rugged engineer, was increasingly necessary for recruiting in an age when both types were almost as extinct as the dodo.

IF THIS GOES ON . . .

With the growing normalization of the industrial research scientist working for a large-scale corporate organization, the romance of the Edison cult and the roughneck engineer lived on in popular mythology rather than in industrial reality. Even so, these cults were still usefully employed for the purposes of recruitment and popular legitimation—especially in the Gernsbackian pages of the pulp SF magazines, where the practical imagination of the boy-inventor wizard could always be relied upon to save the world.[31] One classic example from a 1938 issue of *Astounding* was Jack Williamson's serial story, "The Legion Of Time," in which a choice between alternative futures, good and evil, depends on a boy in a meadow picking up a magnet. That he does so, and is thereby encouraged to become a scientist and an inventor of "dynatomic tensors"—atomic technologies for benign uses—saves the world from the less savory alternative future in which he ignores the magnet to become a "shiftless migratory worker," and in which his redemptive technology is invented and put to despotic ends by less honorable foreigners. In Williamson's story, the utopian city of the good future carries all of the classic architectural features of the futurist vision, hatched in the City Beautiful movements of the Progressive era:

> The ship was two miles high. Yet, so far as his eye could reach in every direction, stretched that metropolis of futurity. Mirror-faced with polished metal, the majestic buildings were more inspiring than cathedrals in their soaring grace. With a pleasing lack of regularity, they stood far apart all across the green parklike valley of a broad placid river, and crowned the wooded hills beyond. Wide traffic viaducts, many-levelled, flowed among them, busy with strange, bright vehicles. Coming and going above the towers, great silver teardrops swam through the air about the ship.[32]

The utopian city of the good future, named Jonbar after John Barr, the boy-inventor, is dominated by a huge statue of Wil McLan, the mathematician who figures prominently in the series as the heroic inventor of an atomic-powered timeship that allows the adventurers to navigate between the alternative futures. The scientifically administered future city honors these rugged individuals from the past in various memorial forms and buildings which bear their name. The scientists of the good future are strictly technocratic, "brisk and efficient" and as undifferentiated as their correlates in the alternative evil future of the foreign-sounding Gyronchi, where an army of giant, hypertrophied ants police a city of flimsy huts dominated by a castellated fortress of red metal on one hill, and a black colonnaded temple on the other. In this respect, one could say that the good far future more clearly resembles the scientific milieu of Williamson's present, honorifically organized around corporations bearing the names of their inventor–founders.

What can we say about the relation between such stories and the socioindustrial reality of science that I have sketched out? It would be critically reductive to explain such stories as simply anachronistic, as symptoms of a lazy, determinist lag between the soloist culture of boy's adventure fiction and the corporate world of social and industrial reality, and that the fiction would eventually catch up. Equally, it is too facile to see the world-saving genius–inventor type as a critical response, however politically unconscious, to monopoly capitalism's incorporation of science. After all, almost all popular fiction (in contrast to middlebrow and highbrow fiction) depends on the narrative vehicle of strong audience identification with individual character types, while its ideological appeal often rests on nostalgia for traditional, or mythical, forms of knowledge and social action. But SF, especially in the thirties, made a special case for itself as an "advanced" genre of popular entertainment, concerned with new, cutting-edge, even prophetic forms of knowledge and social action in the present and in the future. Consequently, the contradictions it displayed as a bearer of the new technocratic ideology had a claim on modernity that other generic popular fiction was not in a position to share or to match.

The survival of the erector-set-inspired amateur inventor in thirties SF, even though it was anachronistic, meant that this heroic figure became available as a protopolitical vehicle. Not unlike the celebrated "small guy" in Frank Capra's films of the time, the figure was used to express a defiant alternative to corporate labor's stifling assembly-line spirit. In both cases, Capra's and Gernsback's, the appeal was ostensibly populist—and ultimately, in Capra's case at least, anti-fascist, in keeping with the spirit of the Popular Front in the mid to late thirties. In an age of high anxiety about technological unemployment, the inventor's autonomy over the creative use of gadgetry was an attractive alternative to the feeling of loss of mastery over technology to the new corporate technostructure. This feeling extended from the shopfloor, where the skills embodied in workers' rules of thumb had been coopted by the new Taylorist managers, to the cockpits, if we believe their complaints, of the old-style captains of industry.

So too, if we take Gernsback's agenda of recruitment literally, we can see that a diet of pulp SF may have been an appealing advertisement for the social virtues of doing science, but it was a poor preparation for likely industrial draftees. Youthful readers, hopes flattered by the Gernsbackian focus on the more attractive and creative elements of the new technocratic ideology, had every chance of seeing those hopes dashed in the employment market and labor conditions of the time. After being raised on a diet of "astonishing," "amazing," and "wondrous" adventures through scientific endeavor, the prospects of being thrilled by the drudgery of sublunary employment in the everyday factories of science and technology were very slim indeed. In a similar way, the adventures of time travel and space travel, the standard imperialistic components of pulp SF, might also be seen as utopian versions of the desire to escape the new Taylorist tyranny of organized and quantified time and space that had come to preside over the contemporary workplace.

Like the genius–inventor, the cowboy and the private detective were anachronistic heroes increasingly employed to criticize the loss of individual autonomy in a bureaucratically organized corporate culture. For the most part, however, the outcome of this romantic individualism was a libertarian critique, usually with conservative implications. This tendency is quite evident, for example, in the early science fiction of Robert Heinlein, which expressed the hatred of state and monopoly capitalism through nostalgia for the free-enterprise frontier of nineteenth-century North America, where invention and rugged individualism are prized. In stories like "If This Goes On . . . " (1940) and "Logic of Empire" (1941), dystopian pictures of future monopoly states—capitalist and theocratic respectively—are challenged by the heroic activities of patriarch–entrepreneurs.[33] Over ten years before, in the heyday of Gernsbackian pulp, similar dystopian stories by David Keller about the loss of autonomy were appearing in *Amazing*: "The Revolt of the Pedestrians" (February 1928), "A Biological Experiment" (June 1928), "The Menace" (June 1928), "The Psychophonic Nurse" (November 1928), and "The Threat of the Robot" (June 1929) were all published in the late twenties.

Towards the end of the thirties, elements of anti-Gernsbackian dystopian thinking were becoming a permanent presence in the pulps, especially in *Astounding*. At the same time, the dreamy rhetoric of technological futurism had been taken over lock, stock, and barrel by corporate advertisers and managers in the business of selling tomorrow's streamlined world. This rhetoric reached its culmination at the 1939 New York World's Fair, billed as "the first fair in history ever to focus entirely on the future," although, by 1939, the tradition of futuristic presentations at such fairs was long established. The Fair, constructed on a reclaimed ash dump in Queens, was one of the largest civil engineering feats of the decade. Initially conceived by a group of businesspeople as a post-Depression stimulant, it was thematically planned to promote the social potential of science by progressive designers and architects like Walter Teague and Robert Kohn. In fact, the designs and theme of the Fair were the high-minded outcome of a debate between traditionalists and "functionalist" progressives, including, most prominently, Lewis Mumford, who argued persuasively for a totally planned Fair devoted to exhibiting the social and historical significance of technological change.

The principal elements of the Fair's philosophy included technology's potential to create a postscarcity culture out of machine rather than human labor; the need to preserve democratic institutions in the face of fascism; and the capacity of social and urban planning to resolve the alienation of people from a communitarian life.[34] In contrast to recently staged fairs in Chicago (1933) and San Francisco (1939), which had sensationalized the spectacle of mechanical progress, the planners of the New York Fair rejected what they saw as the showy appeal of gadgetry for gadgetry's sake. Instead, the Fair showcased entirely planned and integrated living environments—in other words, an attainable near future. While the machine remained the central motive force for social change, the package also highlighted science, art, education, consumer abundance, and environmental wellbeing. All the great designers and architects of Streamline

Moderne were recruited to translate these themes into a dramatic visual language: Raymond Loewy, Norman Bel Geddes, Henry Dreyfuss, Egmond Arens, Donald Deskey, Gilbert Rohde. The Fair became the first showcase for industrial designers, the artist–engineers of the day.

As a result of this lofty debate among intellectuals about the Fair's social themes, which were integrated, wherever possible, into the focal exhibits, a good deal of popular suspicion arose about the high-minded paternalism of this "World of Tomorrow." In response, much of the educational apparatus with its accompanying technological languages was removed in an attempt to make the Fair more people-friendly, and more sensationalist appeal was added in its second, and last, money-losing year in 1940. It didn't help matters that Roosevelt's opening speech underlined the daunting link between the nationalistic pioneerism of the previous century and the new manifest destiny of an extraterrestrial tomorrow when he proclaimed, with Emersonian echoes, that "our wagon is still hitched to a star." In fact, those parts of the Fair that looked like "no place on earth" can be seen, in retrospect, as a semiofficial response to the embryonic cult of the space age launched in the name of science. While the embodiment of the Fair's social philosophy, a streamlined democracy, could be found centralized in the Democracity model inside Dreyfuss's Perisphere, the crowd-pleasers were all located in the Transportation Zone, where the Fair's big commercial sponsors presented the spectacle of corporate SF in practice. It was there, in the Rocketport inside the Chrysler Building (Loewy), on Ford's Road of Tomorrow (Teague) and, above all, in the Futurama of General Motors' Highways and Horizons Pavilion (Geddes), that the gleaming visions of the Gernsback Continuum—the streamlined rockets, teardrop cars, and sleek superhighways—took on their most concrete corporate forms.[35] The Time Capsule that the Westinghouse Corporation buried at the Fair in Flushing Meadow even contained an issue of *Amazing Stories* on microfilm. With great ceremony, it was decreed that the capsule, ominously shaped like a double-headed bullet, would lie unopened until 6939 (like the nuclear waste that the military–industrial complex would soon be burying for just as long), a date that could only have significance in an SF story of the day. Highbrow science, in the form of a letter penned by Einstein, also had its say. In his letter, Einstein laid out the basic dialectic of the technocratic argument, sounding the Panglossian potential of science and technology while lambasting the irrational restraints and vagaries of capitalist production:

> Our time is rich in inventive minds, the inventions of which could facilitate our lives considerably. We are crossing the seas by power and utilize power also in order to relieve humanity from all tiring muscular work. We have learned to fly and we are able to send messages and news without any difficulty over the entire world through electric waves.
>
> However, the production and distribution of commodities is entirely unorganized, so that everybody must live in fear of being eliminated from the economic cycle, in this way suffering for the want of everything. Furthermore, people living in different countries kill each other at irregular time intervals, so that also for this reason anyone who thinks about the future must live in fear and terror. This is due to the fact that the intelligence and character of the masses are incomparably lower than the intelligence and character of the few who produce something valuable for the community.[36]

In these last comments, and in his concluding "trust that posterity will read these statements with a feeling of proud and justified superiority," Einstein was affirming the elitist and frankly undemocratic world-view of a scientific community that had only recently moved into the sphere of social and political activism.[37] In the late thirties scientists had rallied to the cause of anti-fascism, while many had earlier responded to the technocratic invitation (and the Soviet "experiment") to apply the "scientific method" to the planning of a more rational society. Technocracy went against the grain of the presiding disciplinary ideology of the natural sciences—that of Darwinist laissez faire-ism. As Einstein's comments show, however, scientists' autonomous responsibility for planning was unlikely to provide for any democratic decision-making that would involve the "incomparably lower . . . intelligence and character of the masses": planning would be centralized in the hands of experts who could best decide the shape of the future on their own.

Depending on your point of view, Einstein's 1939 vision of the future autonomous rule of intelligence was either premature or already outdated by increasing corporate control over the use of "science" for forecasting and planning the future. At the World's Fair, visitors to the Futurama building, which housed a giant scale model of the United States in 1960, were given an "I Have Seen the Future" button, courtesy of General Motors, on exiting the building. The emphasis was clearly on the spectatorship. As Bruce Franklin points out: "General Motors had constructed a model of America's future. It is the corporation that plans and builds, while the people are purely passive, comfortably watching the creation in motion as mere spectators."[38] In contrast to the polluting automobile hell that lay in GM's real American future, Franklin cites the ecological example of the electric intramural railway, which had been prominently displayed in the name of the future at the Columbian Exposition of 1893. Such a railway was running cheaply and cleanly in all major US cities (especially Los Angeles) by 1935. The purchasing and scrapping of the electric railway system by GM and allied interests in the subsequent years turned out to be one of the century's great crime stories.

In the decade before the New York World's Fair, SF had indeed been part of the cultural process of popular consent that helped to grant the autonomy commanded by these corporate forms, and that ultimately endowed the likes of GM with powers that soon came to preside without ecological foresight over the nuclear militarization of aerospace and the carbon-intensive automobilization of ground space. But it seems quite reductive to hold pulp SF's aesthetic of "progress" to account for its apparent complicity in the corporate version of the Gernsback Continuum. The history of the genre formation of SF, as I have briefly described it here, was certainly intimate with dominant capitalist ideas about science and technology. But it was also responsive to "amazing" ideas about the future of science and technology that went well beyond the limited purview of industrial capitalism, stretching those limits into unmanageable realms of social invention that could never possibly be met by the subsequently deflationary reality of everyday technology. Once it has abolished utopias by announcing the end of ideology, corporate technocracy has to deliver what it promises—incremental raises in consumer gratification—or it is found wanting. SF culture is not part of that risky game. Its futures provide ample room for alternative forms of gratification. Even in those early years, when SF most embodied the technocratic spirit, there was a close link to what I have described as critical technocracy, an attempt, in its heyday at least, to *change* the rules of the game that have governed GM's idea of technological progress.

The exploitationist side of the World of Tomorrow capitalized on the hopes of a traumatized population slowly coming out of the grip of the Depression. But the other side of the Gernsback Continuum, which I have tried to describe here, was complicit with three decades of progressive thinking about technology's capacity to weld together the future and progress into a single, social shape. The aesthetic form of this continuum between future and progress had found its most visible expression in streamlined industrial design—the smooth dynamics of an inevitable horizontal movement of energy, insistently fluid, with no obstacles in its rounded path towards a future. Friction-free, energy-efficient, and seductively constructed around the attractive surfaces of Bakelite, Vitrolite, and newly synthesized plastics, this representation of the hygienic speed of tomorrow became the visual language of progress in the thirties; a sign that the future, to cite a favorite streamline pun, was just "rounding the corner." Pulp SF, which boasted the utopian, teardrop look on every magazine cover, was one of the more popular versions of an aesthetic that signified a genre going somewhere fast. Like the streamline designers, who thought that basic units like the teardrop were Platonic forms, essentially perfect solutions to all design problems in the future, the Gernsbackian version of futurism was untroubled by its ideological assumptions about the future's unilinear shape. Both aesthetics would fall victim to the new logics of obsolescence—social and stylistic—with which *art moderne* was industrially associated. The "future" look would soon be out of fashion, proving, perhaps, that the future really was a continuum, and illustrating one of those time paradoxes of which SF is so fond. The Gernsback Continuum would ultimately leave Gernsbackianism behind.

Of course, there were reasons other than the innovations of industrial design and fashion for the obsolescence of futurism. War and Hiroshima, in particular, gave the future a bad name. But while it lasted, and until it was hired to sell the corporate definition of the American way of life at the World's Fair, the streamlined Gernsback future had been a "natural" expression of progressive thinking about a better society.

At a time when science and technology were becoming the primary rationales for capitalist growth, technocrats, socialists, and progressives each assumed, in a publicly visible way, that they were the historical heirs to a tradition of technological futurism—a tradition not at all adequately described by today's derogatory term "technophilia". For technocrats, it was a tradition in which expertise, rationality, and knowledge challenged the arbitrary diktat of capital; for socialists, it was a tradition in which the technological forces of production undermined the existing social order even as they reinforced it; and for progressives, it was a tradition in which technology was the ally of democratization and the enemy of limited production for profit.

It is not fair to assume, from the vantage-point of today (when Gernsbackian stories about the redemptive wonders and liberatory power of high technology are presented, just as unfaithfully, in the name of ecologically "clean" futures), that the heirs to these traditions were simply not ecologically minded. Exemplary thinkers like Mumford, who had a typically Gernsbackian gadgeteer youth, insisted on seeing technics in an ecological context. Mumford drew not only upon the conservation movement of Progressivism (the result of a deal struck by big business and big government to boost efficiency but which nonetheless introduced environmentalism as a political idea, and which otherwise produced such ultimately fine critiques as Stuart Chase's *The Tragedy of Waste* (1925)), but also on the garden city and regional planning movements that leaned towards the decentralized pastoralism preached by the Russian anarchist Pyotr Kropotkin and the Scottish utopian Patrick Geddes. In *Technics and Civilization* (1934), Mumford produced the most representative document of the progressive side of 1930s technocratic humanist thought. In that grand survey of Western technological cultures, he lays out his qualified faith in the cleanliness and efficiency of the new unseen energies that power the electronics- and physics-based technologies; the task of the "new technics," he wrote, "becomes the removal of the blighted paleotechnic environment, and the re-education of its victims to a more vital regimen of working and living."[39] With the harnessing of these new energies, the machine would become an "ally" of holistic, as opposed to mechanistic, life: "Efficiency must begin with the utilization of the whole man; and efforts to increase mechanical performance must cease when the balance of the whole man is threatened."[40] In his attention to "the calculus of life" over and above the "calculus of energies," and in his ecological rejection of the almost universally admired advances of the Soviet planned economy's worship of mechanical scale and giant production, Mumford embodied the green side of thirties "technophilic" futurism that was often lacking in engineers' and planners' visions of rational production. And in his critical challenge to the "obsolete" structures of profiteering and militarism that shape the research and development of technologies and limit their use to market stimulation or to barbarism, Mumford demonstrates Marx's lessons about a capitalist system that could no more afford full-blown technical progress than it could afford to stagnate without it.

Like even the most skeptical subscribers to the idea of a technological sublime, Mumford, in the 1930s at least (*The Myth of the Machine*, written much later, was a bitter jeremiad, as pessimistic as his technocratic writing had been optimistic), was still placing his faith in the future. He hoped that what he called "geotechnic" and "biotechnic" progress would bring even more "life-sensitive" technologies into a world restored to a state of decentralized pastoral idealism. In this, perhaps, he reaffirmed a pervasive North American ideological vision of what Leo Marx called the "machine in the garden," in which the promise of decentralized democratic community is often advanced as a cover for ever greater commercial exploitation of physical resources and labor power.[41] In this persistent myth, versions of which are shared by left and right like, cultural power is indeed concentrated in nostalgia for a future that will never be. This is the myth of the future that cyberpunk, for example, claims to have forgotten in the interests of the new dystopian realism.

There is little to be gained, finally, from using our hindsight simply to excoriate the "wrongheadedness" of 1930s progressive thinking about technology's capacity to manufacture a better social future. Like the streamlined design, which took on the "natural" look of progress, the codes of technological futurism were "second nature" for the progressive thought of the time. The lesson that science and technology are ideologies in their own right, intimately tied up with bureaucratic organization of power and with domination over nature, had not yet been fully borne out. Futuristic planning on the part of technocratic elites had not yet become fully institutionalized to the point of excluding democratic decision-making procedures. Efficiency was still a matter of public interest and policy, long before it became a byword for privatization. And it would be decades before the ecological specter of dwindling natural resources and global degradation assumed a cogent political form. If there are lessons to be learned from looking back at the now distant shape of these outdated futures, then they must be properly historical lessons.

But another responsibility is called up by any survey of the naivety of historical futurism, and that is the responsibility, examined in the following two chapters, to recognize the naivety of the prevailing modes of futurism today. Again, I would argue that this is a responsibility quite specific to SF, as a popular genre that has learned to ask very serious questions about possible, probable, or preferable futures. In popular culture today, the period "look" of the future is a *survivalist* one, governed by the dark imagination of technological dystopias. Cyberpunk literature, film, and television express all too well the current tendency to unhitch the wagon from the star, to disconnect technological development from any notion of a progressive future. In doing so, they leave the future open to those for whom that connection was and still is a very profitable idea. Once considered the home of progressive thought, the "future," as I will show in the following chapters, has been occupied by corporate and military interests for most of the postwar period. We can no more afford to see ourselves as unavoidably victims of technological development than as happy beneficiaries of a future that has already been planned and exploited. Such an attitude does not lead to empowerment. While it may offer a way out of what Gibson called the "Gernsback Continuum," it is a one-way ticket to a future that we must try to make obsolescent as quickly as possible.

CHAPTER FOUR

CYBERPUNK IN BOYSTOWN

What happened to the World of Tomorrow? Vast sectors of the scientific and technological research needed to deliver the democratic utopia pledged to an expectant populace at the 1939 New York World's Fair had become restricted to military operations just a decade later. Much of the promised research and development in transportation, communications, and new technologies, electronic and atomic, were subordinated to the permanent war economy for the next forty years.

In other respects, the domestic landscape had been all too successfully reshaped in accordance with the vision of 1939. By the time of the second New York World's Fair in 1964, macro-planning and macro-engineering had successfully contrived to rearrange neighborhoods and suburban developments around the central principle of living with the automobile.[1] The appointment of powerbroker Robert Moses as President of the 1964 Fair was a corporate tribute to his notorious career as the engineer of the metropolitan New York postwar environments, wrenched into shape at great ecological and communitarian costs. The Ford, Chrysler, and General Motors buildings at the Fair paid their own tributes to the future of such enterprises. In the GM pavilion, the successor to Futurama, visitors were given the chance to ride across "remote" landscapes, hitherto uncolonized by industry, and to see the "machines of tomorrow" shaping the surface of the moon, the ocean floor, the Antarctic, the jungle, and the desert with the impress of technological development. In the jungle section of the ride, a massive road-building machine, a "factory on wheels," preceded by laser-wielding, tree-clearing machines, was, as GM put it, "capable of producing from within itself one mile of four-lane elevated superhighway every hour." This monstrous parody of every developer's dream was the most eulogistic tribute of all to the Age of Moses. Several pavilions at the Fair were also marked by the spirit and presence of Walt Disney, Moses's other great conspirator in the assassination of nature, and architect, ultimately, of the Epcot Center, the most fully administered of corporate futurist environments. Commentators did not have to look far to find an allegorical origin for the Fair's stories of ecocide; Moses's original "reclamation" of Flushing Meadow for the site of the Fair fitted the bill exactly. As a recent critic put it:

> "Futurama" literature repeatedly refers to the conquest of barren, useless, "waste" land, of making such areas productive, and the like. It is now realized that such lands provide important, often irreplaceable ecological services. The same change in attitude applies to the Fair site itself. It was often disparaged as a swamp, a salt bog, a dump, of no value; the site preparation for 1939 was a major land reclamation effort. Today the cry might very well be one for *restoring* the area as a tidal wetland. But first the groundwater would have to be checked for contaminants from Fishhook McCarthy's old dump.[2]

Although the United States in 1964 was at the height of its postwar boom, in love with the Space Age, and fully subscribed to President Kennedy's New Frontier of science and technology, the World Fair's generic language of progress did not hold the decisive rhetorical sway it had enjoyed in the post-Depression years of the late thirties. The resurgence of the cult of science and invention in the post-Sputnik years did not establish the same deep roots in popular consciousness as it had done in the decade before Hiroshima. Decisive environmental disasters still lay in the near future: Love Canal, acid rain, Three-Mile Island/Chernobyl, Agent Orange, Bhopal, Exxon Valdez, the rainforests, the Gulf War, and global warming. But the social pathology of Bomb culture had too pervasively defined people's horizon of expectations about the world of tomorrow for the rhetoric of unbounded progress to enjoy another round of popular acclaim in the old form of macro-industrial engineering.

The theme of the 1964 Fair, "Peace Through Understanding," struck a hollow note in the semi-alert condition of Cold War. Even more fraught with irony was the Fair's underlying globalist theme of describing a "shrinking world," the term seized upon by the emergent telecommunications giants to advertise the first generation of high-tech tools for transnational empire-building. "Shrinking" the world was an idea with ominous overtones; its embryonic appeal to McLuhan's vision of a global village was offset by its semantic counter-warning about the dwindling of finite resources and the denaturing effects of technological degradation. So too, when the public came to visit the Fair this time, they brought along worldly demands that breached the quarantine space maintained by the Fair's planners. Civil rights protests about the racial policies not only of the Fair itself, but of many of its corporate exhibitors and participant nation-states, were prominent in the opening days.[3] Fierce controversies were generated around the political nature of many of the commissioned art exhibits, and as the air over North Vietnam thickened daily with the new hard rain, the emergent anti-war movement increasingly compromised the Fair's official pacifist theme. Any number of the futuristic projections boldly displayed inside the Fair unconsciously embodied early-warning signs about the coming challenges of the new social movements. Who knows, for example, how many of the teenage, baby-boom girls who visited the General Motors pavilion were confused, perhaps even repelled, by GM's display of the Runabout, an experimental three-wheeled vehicle with a built-in removable shopping cart, designed to cater to modern women's status as "poor drivers" and "avid shoppers"?

While the launch of Sputnik in 1957 had provided the chief stimulus for reviving the American cult of science and invention, the official Western picture of Soviet technology was dominated by the yearly ritual of the May Day parade of military hardware trundling ceremoniously through grey Moscow streets past the assembled ranks of the Politburo. Events like the World's Fair could thus be posed as alternative Western responses to this grim, militaristic image of socialist technology-worship. Such parades, however, would have been much more appropriate if they had taken place in the streets of Washington, at the heart of a political culture that only paid lip-service to the democratic ideals of public accountability. No comparably graphic Western images existed of the massive investment of our socioeconomic resources in such military technology. The wholesale underpinnings of the permanent war economy would only be partially, and fleetingly, revealed by the anti-war movement, and by the opportunistic role played by media exposés of the technological atrocities visited upon the Vietnamese people (a role assiduously denied the media by the military in the recent Gulf War). The real consequences of the threat of nuclear buildup were otherwise confined to a wholly abstract picture of instantaneous annihilation. US Army ground footage of the real effects of the bomb on the survivors of Hiroshima and Nagasaki was suppressed.[4] Public consumption was limited instead to those aesthetically fascinating images of distant mushroom clouds that so quickly became naturalized as part of the future's domestic fantasy-landscape and that became the focus of black comedy in the closing sequence of the film *Dr Strangelove*.

This is not to say that the abstract, aestheticized imagery had no deep purchase upon popular fears and anxieties about the short-term futures of daily life. The corporate state's attempt to depict a nuclear threat was all too successful in this respect, establishing a pervasive sense of determined pessimism while reinforcing, through the spectator's sheer remoteness from the nuclear images, the belief that decisions about this future were always made elsewhere, by people who lived in a cloud of nebulous reason. By the time of the 1964 Fair, science and engineering, the factories of 1939's World Of Tomorrow, had become testing grounds for a postapocalyptic future whose features had been drawn up in the tireless name of "national security."

Popular consciousness about the future in the Cold War period was marked by a pessimism, however much concocted and controlled by the state, in which expectations had no more staying power than throwaway products. "You go and ask people what they think the future's really going to be like. Half of them will tell you they think they're gonna fry at ground zero. What kind of attitude is that for people to have?"[5] Cyberpunk SF writer Bruce Sterling's flip observation resonates with the perceptions of many fellow Americans who grew up in the years of the Civil Defense initiative, concretely brought home in the domestic mystique of the bomb shelter. The temporal effect of this fear was not due to a recession of the present, of the sort experienced during the Depression; after all, the postwar years were characterized by a long consumer boom. Rather, the growth of that spendthrift culture was fundamentally linked, through the omnipresent nuclear threat, to a recession of the future brought on by the nuclear imaginary of instantaneous (and painless) annihilation. By contrast, the onset, in the public mind, of ecological concerns generated an image of the future in slow decay, where the consumerist byword of disposability would increasingly be seen as complicit with the disastrous exhaustion of natural energy and mineral resources. Time was running out in a way that was radically different from all previous eschatological forecasts about the end of the world. With the growth of public consciousness about environmental degradation, the temporality of the future took on a new dimension. No longer the haven of inevitable progress, and no longer the scene of apocalyptic wipeout, the future was now fraught with complex responsibilities for which no easy or coherent utopian narrative was appropriate.

After official proclamations in the early fifties about the sunny benefits of atomic life, disaster-ridden imagery of a futureless future established some of its deepest roots in popular consciousness and popular culture in the 1960s and 1970s.[6] Science fiction of the period, in particular, shifted away from the streamlined, utopian futures that had prevailed before the war, to dystopian projections of science gone awry. *Dr Strangelove*'s wry scenario of elites planning a post-nuclear afterlife at the bottom of mineshafts exposed the growing popular distrust of technocratic decision-making in the early 1960s that would ultimately fuel anti-nuclear activism in the 1970s. Even the physicist Edward Teller, naked apologist for the nuclear state, was prematurely complaining in 1962: "Today I do not read science fiction. My tastes did not change. Science fiction did. Reflecting the general attitude, the stories used to say 'How wonderful.' Now they say 'How horrible.'"[7] Teller had been the very personification of men's strange love for the theory and practice of military annihilation; yet by the early sixties he had conceded, if in characteristically unimaginative form, that the climate surrounding science had shifted to one of suspicion. In contrast to the Gernsbackian hero–scientist of the thirties, the megalomaniac scientist with Gothic undertones had become a permanent, structural paranoia within the Cold War SF film genre. But while there would always be a "bad guy" scientist to finger for dramatic effect, it was more rare to find a thoroughgoing indictment of the logical organization of science as an industry tied in to military needs.

Teller's distaste for dystopian SF trends not withstanding, the military establishment was conscious of the unofficial role that science fiction generally played in the modern history of futurology by constructing the look and feel of various futures, thinkable and unthinkable. In 1978, for example, the Office for Technology Assessment (OTA) was asked to prepare a report, eventually published as *The Effects of Nuclear War*, to estimate the impact of a limited nuclear conflict on the economy and the surviving population. Finding that empirical data alone could not provide an adequate picture, the OTA commissioned a work of science fiction to round out the relevant future scenario. Published as part of the report, Nan Randall's account of post-holocaust life in Charlottesville, Virginia taps into the powerful Jeffersonian mythology associated with that city. In particular, Randall presents agrarianism, of the sort that Jefferson espoused, as one of the most effective, survivalist ways of arresting society's technological retrogression to a late medieval infrastructure.[8]

In the late seventies, such a commission (combining the military "scenario" with an SF alternative world) was a simple piece of genre hack-work, after decades of near-future fiction in love with post-apocalyptic scenarios where survivors either start again *ex nihilo*, or else reconstruct communitarian life under conditions of technological de-evolution. Naturally, the OTA was interested in appraising the chances for developing sustainable technologies that would be appropriate and useful for survival. A sociologically dense description of such a technoculture would help to naturalize the image of survival under adversity. The fact that the military chose to commission such a work was official recognition of SF's proven capacity to produce survivalist handbooks that fed into familiar North American value-systems of self-reliance, pioneering, and pragmatic *savoir faire*.[9]

Science fiction writers, more than those of any other popular genre, have been passionately concerned about their social responsibility to imagine better futures. For many in the SF subculture, this sense of utopian responsibility was slowly eroded in the Cold War period by the dominant dystopian and fatalistic visions of nuclear annihilation, which had an especially powerful influence over the genre's tradition of extrapolating the future. With the appearance of the New Wave movement in the early sixties, writers like J.G. Ballard, Brian Aldiss, Michael Moorcock, Harlan Ellison, Thomas Disch, and Roger Zelany produced a kind of sophisticated, "literary" science fiction that opened up a space for exploring traditional SF genres in a more self-critical way.

By the mid-seventies, writers more attentive to questions of gender and sexuality such as Samuel Delany, Ursula LeGuin, Joanna Russ, Marge Piercy, Sally Miller Gearhart, and Suzy McKee Charnas responded more fully to the New Wave challenge by publishing utopian novels that extended the countercultural critique of scientific rationality into a reexamination of the form of the utopian genre itself.[10] In these novels it was no longer the *content* of utopian thinking that was being reformulated; what they also scrutinized was the formal processes of rendering utopian desires and passions into imaginative shapes. Whereas most earlier utopias had been based on unilinear blueprints drawn up along systematically rational lines, these novels recognized that the task of living differently and living with difference required imagery, action, and decision-making that was more open and heuristic, accommodating ambiguity, uncertainty, and self-criticism. Utopia resided more in personal education or consciousness-raising than in the contours of a perfectly planned society. In this respect, the critical utopias of these writers embodied the critique of technocratic decision-making (rationalist solutions imposed by a consensus of expert elites) that lay at the core of the New Left's advocacy of participatory democracy, while reconstructing and reinventing the future in accordance with the radically different lines of power and desire that were espoused by the more utopian segments of the 1960s counterculture.

With the post-sixties emergence of the new social movements, the universalist basis of traditional utopian thinking in politics lost much of its cultural power, just as the genre of utopian writing imploded under the pressure of its own self-criticism. Critics of science fiction like Fredric Jameson lament the loss of this power, connecting it to our growing "inability to imagine the future" and to the general "waning" of a sense of "historicity" in our culture.[11] But this "waning" can also be interpreted more optimistically in terms of the conditions of its *emergence*—the diverse challenges, on the part of women, sexual minorities, and people of color, to a universalist interpretation of history that are the preconditions for the task of living differently today. In this light, Jameson's story of decline might be seen instead as the story of the atrophy of a particular kind of totalizing historical imagination, exercised in the name of universality.

If utopianism draws its appeal from perceived deficiencies of the present, then the power of dystopian thinking lies in its perceptions about deficiencies of the future. In spite of the nuclear "threat" constructed and administered by the state, SF's dystopian pictures of near-futures and far-futures embodied an important challenge to the dominant culture of the postwar years. But these dystopias lost something of their oppositional quality after the energy crisis of the mid-seventies when images of a dark, eco-dystopian future became the official "look" of the future in popular culture. From the mid-seventies through the eighties, such images pervaded literature, television, music videos, advertising, as well as films like *Escape from New York*, *Logan's Run*, *A Boy and His Dog*, *Soylent Green*, the *Mad Max* trilogy, *Blade Runner*, *The Running Man*, *The Terminator*, *Robocop*, *Aliens*, *Cherry 2000*, *Max Headroom*, *Millennium*, *Brazil*, *Hardware*, and a host of others. The dark scenarios associated with this look arguably carried more cultural power than the nostalgic theme-park constructions of the "Reaganite" *Star Wars* genre.[12]

In the wake of punk culture's brilliant anti-utopian influence, launched by the generation with "no future," the entropic, post-apocalyptic, ragtrade look—layered, makeshift, no-color, and all-purpose—had its moment in high art-fashion, before coming to prevail over hundreds of heavy metal music videos that cast predominantly white rock stars (from Billy Idol to Whitesnake) as rebel survivors in trashed-out urban backdrops. These backdrops were presented as futuristic, although their existing prototype could be found in any inner-city environment, populated for the most part by non-whites.

As an antidote to these suburban fantasies of metropolitan life at its very worst, we might consider the "look" of the hip-hop video. In contrast to the white rockers' taste for urban detritus, the hip-hop aesthetic is devoted to bringing color, style, and movement into inner-city environments, transforming bleak backdrops by graffiti that speaks to the act of creative landscaping, rather than urban decay. Hip-hop's meteoric ascendancy within commercial youth culture represents a genuine movement away from the stylized apathy of the post-punk climate in which the survivalist look had established itself as a dominant expression. As its savvy politics of style and its lyrical commentary become mainstream, hip-hop's leading musicians and producers have emerged as important organic intellectuals within the black community. Politically informed debates about old tensions between expressions of militancy on the one hand, and cultural pride on the other, have resurfaced in a medium that currently lies at the very center of popular culture and popular consciousness.

The black urban community feels the threat of the present more acutely than the threat of the future, and yet it is in black urban youth culture that the creative resources for actively countering the culture of futuristic pessimism have been most successfully generated. What seems most significant about this development is that it owes virtually nothing to traditional projections of the future. On the contrary, hip-hop's success has rested on young blacks' need to renew creatively their culture in the present, by rejuvenating the shared histories of a consistently repressed African–American past. In this respect, hip-hop, which shares the post-modernist sensibility of eclectic appropriation through technosampling and the like, is perhaps the best rebuttal of the argument that postmodern culture depends on erasing the lessons and the materiality of history; and one of our best reminders, despite cyberpunk's claim to the contrary, that alternative cultures cannot be founded simply on futurist principles without any attention to the past.

BOYSTOWN

In the eighties, the most fully delineated urban fantasies of white male folklore were to be found in a series of novels by writers loosely grouped under the name "cyberpunk": William Gibson, Bruce Sterling, John Shirley, Lewis Shiner, and Rudy Rucker (the expanded circle might include Greg Bear, Richard Kadrey, James Patrick Kelly, Walter John Williams, Paul Di Fillipo, Pat Cadigan, Marc Laidlaw, Lucius Sheperd). Istvan Csicsery-Ronay has gone so far as to describe cyberpunk as "the vanguard white male art of the age," for its resexing of the "neutered" hacker in the form of the high-tech hipster rebel who figures as the hard-boiled protagonist in many cyberpunk narratives.[13] One barely needs to scratch the surface of the cyberpunk genre, no matter how maturely sketched out, to expose a baroque edifice of adolescent male fantasies. Here, for example, is Rudy Rucker, tenured professor of mathematics and computer science, SF writer, and author of sophisticated works of non-fiction like *Infinity and the Mind* and *Mind Tools*:

> For me, the best thing about cyberpunk is that it taught me how to enjoy shopping malls, which used to terrify me. Now I just pretend that the whole thing is two miles below the Moon's surface, and that half the people's right-brains have been eaten by roboticized steel rats. And suddenly it's *interesting* again.[14]

Where does this shard of twisted suburban wit come from? Is it a belated symptom of the North American punk sensibility? The negationist fantasy of class-conscious male privilege? Or the self-projection of some repressed Schreberian desire to terrorize the socialized body? Nothing, it would seem, could be further from the polymorphous, ecotopian fantasies that had prevailed in New Wave writing, which cyberpunk rejected as "wet," "hippy," and "utopian."

If punk culture was one of the decisive intervening factors between New Wave and cyberpunk, as Sterling (the movement's chief spokesman) and others have claimed, then this transition was part of the remasculinized landscape of anarcho-libertarian youth culture in the 1980s. Outside of its art-rock orgins in the downtown Manhattan club scene, the punk moment in the US (British punk culture was another story) offered an image-repertoire of urban culture in postindustrial decay for white suburban youths whose lives and environs were quite remote from daily contact with the Darwinist street sensibility of "de-evolved" city life. It is perhaps no coincidence that none of the major cyberpunk writers were city-bred, although their work feeds off the phantasmatic street diet of Hobbesian lawlessness and the aesthetic of detritus that is assumed to pervade the hollowed-out core of the great metropolitan centers. This urban fantasy, however countercultural its claims and potential effects, shared the dominant, white middle-class conception of inner-city life. In this respect, the suburban romance of punk, and, subsequently, cyberpunk, fashioned a culture of alienation out of their parents' worst fears about life on the mean streets.

All through the 1980s, this romance ran parallel with the rapid growth of gentrified Yuppie culture in the "abandoned" zones of the inner cities, where the transient thrills of street culture served up an added exotic flavor for the palates of these pioneers in their newly colonized spaces. It was quite fitting, then, that cyberpunk, most notably in Gibson's novels, took as its generic model (to cannibalize and reconstruct in a classic postmodernist way) the atmospheric narratives of hard-boiled detective fiction. In the twenties and thirties, the hard-boiled crime story was the genre best placed to explain the urbanization of North American life. While pulp science fiction of the period was aimed at the stars, the crime story was the urban supplanter of the frontier Western genre, later to make a comeback on film and television on the wave of pastoral nostalgia that accompanied suburbanization in the fifties.[15] Yuppie gentrification was the new pioneer frontier of the 1980s, and cyberpunk was one of its privileged genres, splicing the glamorous, adventurist culture of the high-tech console cowboy with the atmospheric ethic of the alienated street dick whose natural habitat was exclusively concrete and neon, suffused with petrochemical fumes.

In this respect, the story of cyberpunk was a tale about the respective psychogeographies of country (suburb) and city. But its main claim to postmodernity lay in its treatment of the less geographically distinct realm of space and time that was now available through information technologies, the cartographic coordinates of technosimulated space that have no fixed geographic referent in the physical landscape. Here was Gibson's celebrated "cyberspace":

> an abstract representation of the relationships between data systems . . . the colorless nonspace of the simulation matrix, the electronic consensus-hallucination that facilitates the handling and exchange of massive quantities of data . . . mankind's extended electric nervous system, rustling data and credit in the crowded matrix, monochrome nonspace where the only stars are dense concentrations of information, and high above it all burn corporate galaxies and the cold spiral arms of military systems. (*Burning Chrome*, p. 178)[16]

It was in this space, the new "natural" frontier environment for Gibson's console cowboys to roam around, that cyberpunk sketched out the contours of the new maps of power and wealth with which the information economy was colonizing the global landscape. No national frontiers here to control the flow of information, no public or civil space for individuals to access at will, no regulatory body except for the Turing Police, who keep the AIs in check, and the Fission Authority, who police access to and safeguard the security of the corporate databanks. Following Kumiko's adventures in London, in *Mona Lisa Overdrive*, we are even surprised to come across evidence of a country—Britain—that still has a "government" (p. 218).

Cyberspace, and the globally wired, satellite media Net that is a permanent feature of the cyberpunk landscape, is the heady cartographic fantasy of the powerful, aestheticized by Gibson to the point of taking on mystico-metaphysical dimensions. Its ecology of corporate space, neither inner nor outer, is the realm of postmodern *angels,* at least on the old humanist scale of the chain of being. The wealthiest of the corporate clans and magnates "are no longer even remotely human" (*Count Zero*, p. 18), or they are quasi-divine constructs with multiple identities, "ghost[s] called up by the extremes of economics" (*Burning Chrome*, p. 123). The affection with which Gibson lingers over the details of this angelic life, buttressed by the "laws of corporate evolution," is surpassed only by the mysticism he invokes to describe the AI self-consciousness. Even the angels, in this instance Marie-France, matriarch of the Tessier-Ashpool high-orbital clan, lack an adequate Thomist language for the sentient life of artificial intelligence:

> "When the moment came, the bright time, there was absolute unity, one consciousness. But there was the other."
>
> "The other?"
>
> "I speak only of that which I have known. Only the one has known the other, and the one is no more. In the wake of that knowing, the center failed; every fragment rushed away. The fragments sought form, each one, as is the nature of such things. In all the signs your kind has stored against the night, in that situation the paradigms of *vodou* proved most appropriate." (*Mona Lisa Overdrive*, p. 215)

In Gibson's novels, armchair theorists of the self-consciousness of the cyberspace matrix (chronologically achieved in his Sprawl trilogy at the end of *Neuromancer*) speak scholastically of "first causes." For the less well informed, like *Count Zero*'s hacker Bobby Newmark, encounters in cyberspace with otherworldly intelligence are awesome moments of grace: "something *leaned in*, vastness unutterable, from beyond the most distant edge of anything he'd ever known or imagined, and touched him" (*Count Zero*, p. 20).

Such moments of contact with the inhabitants of cyberspace were also part of the postmodern rewriting of the SF tradition of "alien encounters." Here, the fear of unfamiliar, superior intelligence is situated on Earth, within the known parameters of socioeconomic life, everywhere fixed and defined by the equation of knowledge with power. Here, the aliens are both "Us" and "not-Us," evolved hybrids of the corporation as a "life form" that is the "planet's dominant form of intelligence" and whose blood "is information, not people." By contrast, class-consciousness in the lower social world is less equitably sketched out. Gibson's spare perspective on "the masses" makes them barely distinguishable from the beasts and plants at the lower end of the medieval chain of being: "Summer in the Sprawl, the mall crowds swaying like windblown grass, a field of flesh shot through with sudden eddies of need and gratification" (*Neuromancer*, p. 46). Generally, inhabitants of Gibson's Sprawl (the Boston–Atlanta metropolitan strip) are faceless drones, unless they are defiantly marked by membership in the colorful (youth) subcultures like the Lo Teks, the Zionites, the Jack Draculas, the Panther Moderns, the Big Scientists, the Gothicks and the Casuals, whose renegade street knowledge and techno-savvy—"the street finds its own uses for technology"—serves as a social conduit for acts of anarcho-resistance within the interstices of the cyberspace net.

Ultimately, Gibson's chosen playground is the fluid class environment of the mercenary, subcriminal underworld, whose denizens have escaped the fate of early indenture to a corporation only to be caught up as lowly recruits in the game of corporate espionage. The prototype can be found in Gibson's story, "Johnny Mnemonic," whose protagonist is the model for the data thief, Case, in *Neuromancer*. Johnny, whose brain serves as a data storage facility for rent, represents a new form of alienated labor for the information economy. A kind of *idiot-savant*, unable to access the information in his head—"I only sing the song"—he finds that he is functionally incapable of acting upon that information:

> And it came to me that I had no idea at all of what was really happening, or of what was supposed to happen. And that was the nature of my game, because I'd spent most of my life as a blind receptacle to be filled with other people's knowledge and then drained, spouting synthetic languages I'd never understand. A very technical boy. Sure. (*Burning Chrome*, pp. 23–4)

Johnny's consciousness-raising story recounts how he accesses this knowledge, with the aid of an ex-military cyborg dolphin, and becomes "the most technical boy in town," learning all of his clients' trade secrets. This adventurous Bildungsroman is more fully sketched in *Neuromancer*, where the hacking skills of the data thief Case are unwittingly mobilized in the service of the AI Wintermute's attempt to free itself from regulation by its corporate masters. Faced with the task of persuading 3Jane, one of the Tessier-Ashpool clone-daughters, to acquiesce in the AI's plan by releasing a code name, Case's argument runs like this:

> "Give us the fucking code. . . . If you don't, what'll change? What'll ever fucking change for you? You'll wind up like the old man. You'll tear it all down and start building again! You'll build the walls back, tighter and tighter. . . . I got no idea at all what'll happen if Wintermute wins, but it'll *change* something." (*Neuromancer*, p. 260)

By most readers' standards, Case has a limited social imagination, and his life, after all, depends on this argument. Even so, his pleas that *any* kind of change is better than the status quo are rather thin justification for allowing such monstrous intelligences free rein over the information networks. His argument reaffirms the sense that the decisions that count are always being made elsewhere, in circumstances well beyond the control of interested stiffs like Case or even his more ingenious accomplice, Molly Millions. Despite the technical education in the workings of power that they undergo, such people are usually even less in control of their futures at the end of a Gibson adventure than they were to begin with. The same could be said for the star scientists whose innovative research has the potential to shatter knowledge paradigms, revise entire fields, and bankrupt giant corporations. Their ability to discover and develop basic patents—"the high, thin smell of tax-free millions that clung to those two words" (*Burning Chrome*, p. 115)— makes them closely guarded hostages in the intercorporate wars waged through scientific espionage, and hence powerless to act in the public interest.

This tendency has led critics like Peter Fitting and Tom Moylan to complain about the political irresponsibility of Gibson's novels. They harbor no utopian impulses, offer no blueprint for progressive social change, and generally evade the responsibility to imagine futures that will be more democratic than the present.[17] On the other hand, Pam Rosenthal has argued that cyberpunk, like popular culture in general, is usually not the best place to expect to find articulate political directions: "Popular culture," she writes, "promises, at best, to give narrative and symbolic coherence to popular questions and anxieties. It does not promise structural solutions; historical and political analysis and practice—history, in a word—is what's supposed to do that."[18] Rosenthal's may be a timely reminder of the cultural specificity of literary, even popular literary, form. But more than in any other popular genre, the SF community has maintained the demand upon its writers to acknowledge exactly this kind of "responsibility." One of the most familar charges in the science fiction community is that this or that writer is guilty of "celebrating technology" and thus of being politically irresponsible.

Indeed, it was precisely on the point of *responsibility* for the depiction of futures that cyberpunk as a publicized movement made its most forceful claim as new kid on the block. In his introduction to Gibson's *Burning Chrome*, Sterling argued that Gibson's commitment to portray a "credible future" exemplified a responsibility that SF writers, drawn in recent decades to the post-apocalyptic genre, to sword-and-sorcery, and to modern space opera, had "been ducking for years." Sterling argued that this "intellectual failing" to "tangle with a realistic future" was redeemed by Gibson's dedication to depicting "a future that is recognizably and painstakingly drawn from the modern condition." Sterling went on to contrast the Gernsbackian SF scenario—"a white-bread technocrat in his ivory tower, who showers the blessings of superscience upon the hoi polloi"—with the messy street action of the cyberpunk future—"a deranged experiment in social Darwinism, designed by a bored researcher who kept one thumb permanently on the fast-forward button" in which "Big Science . . . is a sheet of mutating radiation pouring through a crowd, a jam-packed Global Bus roaring wildly up an existential slope" (*Burning Chrome*, pp. 2–3). Despite the obvious appeal of Sterling's comparison, there doesn't seem to be much choice here—no chance for a plebiscite on this bored/deranged researcher–designer of a laissez-faire future; no room for other, less frenetic, less perilous models. Sterling's rhetoric cries Wolfe, telling us that the Keseyan Global Bus is what's happening, and so we'd better climb aboard. It's hard not to respond to Sterling's spectacle of SF "lurching from its cave into the bright sunlight of the modern Zeitgeist." But whose zeitgeist are we talking about?

Cyberpunk's "credible" near-futures are recognizably extrapolated from those present trends that reflect the current corporate monopoly on power and wealth: the magnification of the two-tier society, the technocolonization of the body, the escalation of the pace of ecological collapse, and the erosion of civil society, public space, popular democracy, and the labor movement. Cyberpunk's idea of a counterpolitics—youthful male heroes with working-class chips on their shoulders and postmodern biochips in their brains—seems to have little to do with the burgeoning power of the great social movements of our day: feminism, ecology, peace, sexual liberation, and civil rights. Curiously enough, there is virtually no trace of these social movements in this genre's "credible" dark future, despite the claim by Sterling that cyberpunk futures are "recognizably and painstakingly drawn from the modern condition." However modern the zeitgeist of cyberpunk, it was clearly a selective zeitgeist. However coherent its "narrative symbolization" of modern technofuture trends, it was clearly a limited narrative, shaped in very telling ways by white masculinist concerns. And however rebellious its challenge to SF traditions, the wars, within the SF community, between the cyberpunks, the New Wave, and the New Humanists were all played out in boystown.

Consider how the cyberpunk image of the techno-body played into the crisis of masculinity in the eighties. In popular culture at large, symptoms of the newly fortified contours of masculinity could be found in the inflated physiques of Arnold Schwarzenegger and Sylvester Stallone, and a legion of other pumped-up, steroid-fed athletes' bodies. Once described as "condoms stuffed with walnuts," these exaggerated parodies of masculine posture in the age of Reagan were at once a response to the redundancy of working muscle in a postindustrial age, to the technological regime of cyborg masculinity; and, of course, to the general threat of waning patriarchal power. Cyberpunk male bodies, by contrast, held no such guarantee of lasting invulnerability, at least not without prosthetic help: spare, lean, and temporary bodies whose social functionality could only be maintained through the reconstructive aid of a whole range of genetic overhauls and cybernetic enhancements—boosterware, biochip wetware, cyberoptics, bioplastic circuitry, designer drugs, nerve amplifiers, prosthetic limbs and organs, memoryware, neural interface plugs and the like. The body as a switching system, with no purely organic identity to defend or advance, and only further enhancements of technological "edge" to gain in the struggle for competitive advantage. These enhancements and retrofits were technotoys that boys always dreamed of having, but they were also body-altering and castrating in ways that boys always had nightmares about. The new survivalist fantasy of the cyberpunk street guerrilla body would be an expensive one, the consumer mainstay of many a technointensive industry. Such a body would be a battleground in itself, where traditional male "resistance" to domination was uneasily coopted by the cutting-edge logic of new capitalist technologies. But this body was also part of a failing political economy. If the unadorned body fortress of the Rambo/Schwarzenegger physique expressed the anxieties of the dominant male culture, cyberpunk technomasculinity suggested a growing sense of the impotence of straight white males in the countercultures.

A similar story could be told about the hard-boiled narrative chosen as cyberpunk's favored generic vehicle. The adventure formula that Gibson used, and others imitated, offered a pulp narrative that was unable to accommodate the full range of socially critical perspectives on the future that had been present in, say, the feminist utopian SF novel of the seventies. What it did signify, however, were certain defensive characteristics of masculinity in retreat. Nostalgia for hard-boiled masculinity ran high throughout the eighties, especially in the retro-quotationism of the style market. Taking on the cool nihilism of the postpunk mood, while serving as a protective guard against the pervasive jingoism of these years, the hard-boiled style also offered the requisite attitude for the politically cynical struggling to survive in a decade driven by commercial avarice, a decade few people survived with dignity. Deckard's Chandleresque voice-over in *Blade Runner*—wounded, fatalistic, and drenched with distant sentiment—was a stylized icon for such people in this decade.

The hard-boiled narrative of the 1920s and 1930s had been tailored for an urban milieu cheapened by greed and commerce, where personal dignity, romantic love, and asocial desire were outlawed by the dominant institutions. The proto-existentialism of the private eye defined a survivalist sensibility driven underground and into the lonely, lawless environment of the mean streets, where women were more likely to be competing threats than fellow fugitives from the sanctities of family life, and where social Darwinism justified urban poverty and business corruption alike. The hard-boiled urban adventure milieu was a psychogeography of the frontier (like the cyberpunk "interzone"), but claustrophobic rather than expansive like the Western, and thus confined to alienated, anonymous spaces, visually coded in the chiaroscuro expressionism of 1940s film noir.

Gibson's own heavy literary debt to the hard-boiled writing of these decades is easy to see. Metaphors that do special effects in a cheap accent: "getting a bargain from the Finn was like God repealing the law of gravity when you have to carry a heavy suitcase down ten blocks of airport corridor" (*Burning Chrome*, p. 180). Or looser chunks of bar-room maudlin:

> I wasn't happy. I couldn't remember when I had been happy. "You seen your luck around lately?"
>
> He hadn't, but neither had I. We'd both been too busy.
>
> I missed her. Missing her reminded me of my one night in the House of Blue Lights, because I'd gone there out of missing someone else. I'd gotten drunk to begin with, then I'd started hitting Vasopressin inhalers. If your main squeeze has just decided to walk out on you, booze and Vasopressin are the ultimate in masochistic pharmacology; the juice makes you maudlin and the Vasopressin makes you remember, I mean really remember. Clinically they use the stuff to counter sterile amnesia, but the street finds its own uses for things. So I'd bought myself an ultraintense replay of a bad affair; trouble is, you get the bad with the good. Go gunning for transports of animal ecstasy and you get what you said, too, and what she said to that, how she walked away and never looked back. (*Burning Chrome*, p. 195)

And, above all, the lessons learned from the supercharged naturalism of Hammet, whose evocation of sentiment through objects is omnipresent in Gibson's favored lists of industrial detritus, kipple, and "semiotic junk," in his many collections of throwaway objects that "must have been new and shiny once, must have meant something, however briefly, to someone" (*Burning Chrome,* p. 145):

> Turner and Angela Mitchell made their way along the broken sidewalks to Dupont Circle and the station. There were drums in the circle, and someone had lit a trash fire in the giant's marble goblet at the center. Silent figures sat beside spread blankets as they passed, the blankets arrayed with surreal

> assortments of merchandise: the damp-swollen cardboard covers of black plastic audio disks beside battered prosthetic limbs trailing crude nerve-jacks, a dusty glass fishbowl filled with oblong steel dog tags, rubber-banded stacks of faded postcards, cheap Indo trodes still sealed in whole-saler's plastic, mismatched ceramic salt-and-pepper sets, a golf club with a peeling leather grip, Swiss army knives with missing blades, a dented tin wastebasket lithographed with the face of a President whose name Turner could almost remember (Carter? Grosvenor?), fuzzy holograms of the Monument . . . (*Count Zero*, p. 201)

Or here, in an upmarket shopping list:

> A freezer. A fermenter. An incubator. An electrophoresis system with integrated agarose cell and transilluminator. A tissue embedder. A high-performance liquid chromatograph. A flow cytometer. A spectrophotometer. Four gross of borosilicate scintillation vials. A microcentrifuge. And one DNA synthesizer, with in-built computer. Plus software. (*Burning Chrome*, p. 113)

In choice moments, Gibson reduces the naturalist mode to a minimalist shock strategy. Nowhere is this more striking than when the ecosphere is presented as a technosphere, as in the unforgettable opening line of *Neuromancer*—"The sky above the port was the color of television, tuned to a dead channel"—which brazenly announces that henceforth everything here, even the sky, the home of the weather, will be a mediated *second nature*. Three pages later, the aftershock of this "criminal ecology" rolls off the view of Tokyo Bay: "a black expanse where gulls wheeled above drifting shoals of white styrofoam." Characterization often makes use of the same technonaturalism: Molly Millions's body flank has all "the functional elegance of a war plane's fuselage" (*Neuromancer*, p. 44), and the Finn looks as if he has "been designed in a wind tunnel" (*Neuromancer,* p. 48). Libidinal affect is often displaced on to what Gibson calls the "sexuality of junk,"[19] in more direct erotic contact with prosthetic body parts or cyberspace decks, or with the less tangible space of the fiberoptic Net. Desire is learnt through the media, either in the popular *simstim* entertainments or in the spin-off looks of their stars, impressed on their fans' faces through cheap cosmetic surgery.

However self-consciously *literary* (in this, he falls far short of the New Wave writers), Gibson's primary interest lies in recovering the atmospheric eroticism of hard-boiled noir, and, of course, in using the caper narrative, in his words, as a "safety net." His novels are the best evidence that hard-boiled masculinity, second time around, was an appropriate masquerade for a fictional male heroism that chose bad faith as a riposte to the official eighties spectacle of wrapping the male body in Old Glory. The hard-boiled conventions had different political meanings when revived in the 1980s in the lesbian detective novels of Sarah Schulman, Barbara Wilson and Mary Wings.[20]

CYBERPUNK AND DIFFERENCE

Despite the public show of solidarity as a literary "movement," other cyberpunk writers distanced themselves from Gibson's pseudo-mystical devotion to the technological sublime. For example, John Shirley's *Eclipse* novels (1985–88) focused on the *political* rather than the economic shape of their near-future, dominated by international right-wing alliances among the North American Moral Majority, the British National Front, and New Right fascist groups in Europe, all supported by Nato and transnational corporate police forces. A motley group of resistance forces, presided over by the discordant spirit of punk rock, wage a war of position against the anti-terrorists. The resistance cadres appropriate and use technology for their own insurgent purposes.

Bruce Sterling's work is concerned with the *ideological* shape of futures, where entire philosophical systems are formed around new technologies. *Schizmatrix* (1985) recounts the grand galactic conflict between Shaper organicist ideology, whose humanist aristocrats use biological and genetic engineering to prolong life and advance bodily evolution, and Mechanist hard science ideology, whose technocrats have fashioned durable and efficient cyborg bodies for themselves. In the course of the novel, the ancient struggle between the Shaper psychotechnologies and the wirehead pragmatism of the Mechanists is displaced by new galactic ideologies like Zen Serotonin, and other New Age, millenialist philosophies, each of them posthumanist in ways that seal the fate of the older techno-belief systems. A similar concern with techno-ideologies shapes *The Difference Engine* (1990), co-written with Gibson, in which the political philosophy of

Industrial Radicalism has come to parliamentary power in Britain in the mid-1900s on the back of a technological revolution generated by Charles Babbage's invention of the "analytical machine," a mechanical computer.

Sterling's *Islands in the Net* is a twenty-first-century political novel that rejects most of the conventionally dystopian features of the cyberpunk "bad future." Nuclear weapons have been banned, and ecological collapse has been averted. Corporate reform has produced a measure of economic democracy for employees who enjoy access to relatively non-hierarchical decision-making processes. The corporate philosophical model is New Millennium, and the ideology under scrutiny is globalism, technologically promoted by the Net and sustained by its transnational police forces. The political demimonde is located in the renegade data havens of Singapore and Grenada, where libertarian anti-imperialist philosophies thrive, and cash-free economies exist to stave off Third World "sufferation." Africa, excluded from the Net, has become a dump for pre-Millennium technological junk and toxic waste. The novel's character vehicle is a Yuppie female corporate employee, whose business trip to the Net-less Third World leads her into an adventure zone where the international crime of yakuza gangs meets the counterimperialism of insurgent tribal alliances. Despite her exposure to extremes of political violence, in which the forces and allies of the Net are victorious, the lessons she learns about globalism are banal: "One world means there's no place to hide." Her character, however, is pointedly unglamorous by cyberpunk standards and offers Sterling an unconventional vehicle for exploring the contradictions of Net globalism. For the most part she is an unwitting Net agent in the complex struggles in which she becomes involved, and her attempt to make sense of this position is richly ironic but only marginally more satisfying than Case's assessment of his part in the self-liberation of the AI in *Neuromancer*: "She had been part of this, she thought. . . . She had been doing the work of the world—she could sense the subtle flow of its Taoist tides, buoying her up, carrying her."[21]

A similar dearth of human agency pervades Lewis Shiner's *Deserted Cities of the Heart* (1988), which airlifts naive white North Americans into the high Mayan country of Mexico, where US and state counterinsurgency troops are pitted against a rebel guerrilla movement at the height of its rural power. In this case, the forces that come to shape the action are not postmodern but ancient historical ones: aligned with the cosmological temporality of the Mayan calendar, coming to the end of a long phase, and generating cataclysmic events in the natural world. Against this backdrop, the guerrilla war comes to be seen as a trivial skirmish. Mythological history, not human action, is the killing ground and the first cause.

In both these stories—Sterling's and Shiner's—central female characters are especially weakened by their incapacity to act independently of external forces; nonetheless, they make their "intuitive-feminine" peace with these forces through some personal transformation. Lindsey, the office worker who makes do in Shiner's Mexico with a Spanish phrasebook, and who comes to feel the naturally righteous "way" of non-violence. Or Carla, the rebel leader, whose hardened revolutionary zeal is overly tempered, in the view of her male cadres, by her spiritual attachment to the "mystical shit" of the coming Mayan alignment. And Sterling's Yuppie Laura, who estranges her mundane husband and finally becomes attuned to the Taoist tides of corporate power. In their appeal to "alternative" feminine values, these portraits of women who discover a new sense of identity are attempts to represent women outside of the survivalist type of the female cyberpunk "razorgirl" characterized by Gibson's Molly Millions, a hardened techno-altered moll, highly skilled in martial arts and capable of outmatching all her competitors on traditionally masculine terrain.

In the course of the 1980s, variants of the Molly Millions technotype appeared across the whole cultural spectrum, from comic books to avant-garde fiction. At one end was Elektra Assassin, the psychotic, avenging ninja warrior who first appeared in Frank Miller's versions of the *Daredevil* comic and who later emerged as a superhero in her own right in the series *Elektra Assassin* (1986–87), the first celebrated collaboration between Miller and Bill Sienkiewicz.[22] At the other end was Abhor, the cyborg pirate partner of Thivai in Kathy Acker's colorful novel, *Empire of the Senseless* (1988), which features a plagiaristic/piratical commentary on *Neuromancer*. Both characters are steely, orphanesque survivors of a history of victimage that includes paternal rape, followed by repeated sexual predation on the part of violent males. Both play out their adventurist roles in Third World environments—Elektra in Central America, Abhor in revolutionary Algeria—where the sexualized bonds of aggressivity stretched across the economic and racial inequalities created by multinational capitalism. Both preserve their strong status as "free women" at a heavy cost: the incapacity to establish human relationships in a milieu of loveless cruelty, sexual slavery, and addictions to power and wealth that fix women in subjugatory thrall.

The survivalist environment of these "bad" cyborg girls is governed by the rules of a zero-sum game: predator or prey, all or nothing. In Acker's novels, especially, the slaver–victim psychology is employed at every turn as a harshly realistic commentary on a dog-eat-dog world ruled by economic and sexual dependency. Where love is impossible, survival is guaranteed by begging, stealing, and borrowing—a strategy reproduced in the fictional technique that has made Acker the plagiarism queen of modern writing. Libertinism, the other side of Acker's coin, is an equally desperate game, a pleasure pursued in the context of extreme danger and self-interested to the point of rejecting all "correct" utopian appeals towards a collective women's culture based upon "feminist" desires. Exposing and exploring the rituals of power in this way has its undeniable critical rewards. But when it is presented as the only option it also runs the risk of reproducing these same socialized rituals, especially for women who, arguably, have had little to gain from them. In these female versions of the cyberpunk attitude, learning about the future of sexual identity is locked into a restricted set of choices, determined more by the habits of power in the present and the past than by the autonomous capacity of women to feel their way beyond such habits.

A similar example of this structure of limited learning options can be found in the roleplaying game marketed by Talsorian Games as *Cyberpunk: The Roleplaying Game of the Dark Future*, which incorporates many of the settled features of the cyberpunk world, and whose scenarios were written with the help of cyberpunk writers like Walter Jon Williams. In this game, the recognizable future scenario is an extension of the two-tier society created in the last fifteen years between, on the one hand, the buoyant corporate Yuppie sector, and, on the other, the low-wage, service-sector economy with its underside of homelessness, child poverty and drug-assisted destruction of inner city populations.[23] Players of the game are offered a variety of roles—rockerboys (girls), solos, netrunners, techies, medias, cops, corporates, fixers, nomads—while the rules of strategic advantage govern a basic struggle between the rebel alliance (rockers, nomads, solos, and medias) and the corporate forces. The ideological stakes, according to the handbook, are rather vague:

> The traditional concepts of good and evil are replaced by the values of expedience—you do what you have to do to survive. If you can do some good along the way, great. But don't count on it.
>
> *Cyberpunk* characters are survivors in a tough, grim world, faced with life-and-death choices. How they make these choices will have a lot to do with whether they end up as vicious animals roaming a ruined world or retain something of their basic humanity. *Cyberpunk* characters are the heroes of a bad situation, working to make it better (or at least survivable) wherever they can. Whether it takes committing crimes, defying authority, or even outright revolution, the quintessential *Cyberpunk* character is a rebel with a cause. As a *Cyberpunk* roleplayer, it's up to you to find the cause and go to the wall with it.[24]

Even as roleplaying games go, *Cyberpunk* is highly complex. But in a genre that rejects the competitive, win-or-lose structure of the orthodox board game, all of the roles are still governed by the zero-sum action principle of "the cyberpunk way": waste 'em or be wasted. The structure of the game itself can thus be seen as an efficient response to the cyberpunk view of survivalism in a future world where the rules have already been written in the present. True to the adaptational educational thinking from which roleplaying games evolved, the education of desire proceeds through learning and interpreting the rules of play, *not* by changing them.

ALTERED STATES

One of the quandaries faced by players of the *Cyberpunk* game is that the much sought after cybernetic enhancements to the body can only be won at a price. Each piece of metal and plastic added to the body results in an erosion of human identity. Further personality fragmentation and a breakdown of empathy lead to "cyberpsychosis." Behind this idea lies a long history of anxieties about "dehumanization" by technology; a quintessentially humanist point of view which sees technology as an autonomous, runaway force that has come to displace the natural right of individuals to control themselves and their environment. The individual struggles against the forces of the machine for her natural freedom. "Technology" is demonized as if it were inherently oppressive rather than an instrument socially organized for the purpose of domination. Surely this conception of humanism—this desire for a nonalienated self—is an inadequate way of describing a postmodern world of social relations ordered by difference rather than essence, a world in which technology is a mode of social organization where identity is constructed rather than pre-given?

Critics more attentive to these questions of difference and power have seen the technologically colonized (cyborg) body as a new phase of regulatory authority exercised by the corporate state. In her influential essay, "A Manifesto for Cyborgs," Donna Haraway urges a new sense of realism about our cyborg condition, recognizing the new daily sphere of human–machine interface not only as a product of power relations but also as a potential site for contesting and redefining those relations. In rejecting the "naturalist" basis of feminist appeals to the anti-technological, organic wholeness of the body, Haraway calls for ever more transgressive acts in the "border war" between humans and machines. She proposes cyborgism as an imaginative resource or myth for women who are traditionally socialized away from technology and yet who are most often the primary victims of technology in the workplace, the home and the hospital.[25]

Haraway's argument against a common, essentialist identity is post-naturalist rather than, say, postfeminist. In this respect, it takes issue with the humanist critique of freedom that still governs most of the debates about technology's "threat" to the sanctity of the natural, unalienated body. For example, ethical critiques of biotechnology's redefinition of the concept of human life, or the development of artificial intelligence, appeal largely to the humanist faith in this sanctity. Such critiques are less commonly aimed at the powerful corporate–military interests that govern the development of these technologies. For women, who have had little to gain, historically, from the humanist conception of technology as an extension of male freedom to dominate the physical world, there is less cause to observe this creed, and every reason to challenge its concomitant faith—enshrined in certain strains of feminist naturalism—in the organic sanctity of the (female) body. The political importance of Haraway's "blasphemy" lies in her unwillingness to cede the new ground of cyborg power relations by retreating to the mythical space of Edenic naturalism.

This is not say, however, that Haraway's technocultural call for women to resist, transgress, and appropriate goes out on the same wavelength as the cyberpunk call for "using technology before it is used on you." Feminist insistence on the *difference* of bodily relations to technology places limits everywhere on the white-masculinist embrace of the cyborg, however countercultural, that is to be found at the core of the cyberpunk sensibility. Bad white boys, unlike their female counterparts, can draw upon a long history of benign tolerance for their rebel roles, while their male and female counterparts of color are marked as a pathological criminal class. The values of the white male outlaw are often those of the creative maverick universally prized by entrepreneurial or libertarian individualism.

As always, this difference can best be demonstrated by showing how it is elided, in this case, in the universalist name of the "species." The cyberpunk "embrace" has been seen as an emergent stage of "human" development, and is often cited in technocultural contexts that are fully in keeping with the tenets of evolutionary humanism. In certain New Age circles, for example, the radical mutations in bodily ecology imagined in cyberpunk culture are welcomed as an advance in human evolution. The human species is about to get a major upgrade; the cyberpunk will be the inevitable next step in the history of evolutionary forms. Homo Cyberneticus comes after Homo Faber. The purest expression of this philosophy can be found in the magazine *Reality Hackers* (sometime *High Frontiers*, now *Mondo 2000*), which preaches the continuity of today's creative cyberculture with the liberatory individualism of the sixties counterculture. Timothy Leary, the high priest of masterless individualism, is alive and well in a milieu filled with slogans like "Turn On, Boot Up, and Download," "Reversing Entropy is Everybody's Business," or "We Are the New Prometheans. Steal This Fire!" The frontier rhetoric of discovery and creative invention links the LSD spirit of synthetic self-transformation with the technofantasies of cybernetic consciousness:

> We'll bring you the latest in human/technological interactive mutational forms as they happen.

> We're talking Cyber-Chatauqua; bringing cyberculture to the people! Artificial awareness modules. Visual music. Vidscan magazines. Brain-boosting technologies. William Gibson's Cyberspace Matrix—fully realized!
>
> Our scouts are out there on the frontier sniffing the breeze and guess what? All the old war horses are dead. Eco-fundamentalism is out, conspiracy theory is démodé, drugs are obsolete. There's a new whiff of apocalypticism across the land. A general sense that we are living at a very special juncture in the evolution of the species.
>
> Back in the sixties, Carly Simon's brother wrote a book called *What to Do Until the Apocalypse Comes*. It was about going back to the land, growing tubers and soybeans, reading by oil lamps. Finite possibilities and small is beautiful. It was boring!
>
> Yet the pagan innocence and idealism that was the sixties remains and continues to exert its fascination on today's kids. Look at old footage of *Woodstock* and you wonder: where have all those wide-eyed, ecstatic, orgasm-slurping kids gone? They're all across the land, dormant like deeply buried perennials. But their mutated nucleotides have given us a whole new generation of sharpies, mutants and superbrights, and in them we must put our faith—and power.
>
> The cybernet is in place. If fusion *is* real, we'll find out about it fast. The old information elites are crumbling. The kids are at the controls. . . . We're talking about Total Possibilities. Radical assaults on the limits of biology, gravity and time. The end of Artificial Scarcity. The dawn of a new humanism. High-jacking technology for personal empowerment, fun and games. Flexing those synapses! Stoking those neuropeptides! Making Bliss States our normal waking consciousness. *Becoming* the Bionic Angel.[26]

This is the discourse of maverick humanism in full flow, headily pursuing its goal of an "assault on limits" in the name of individual self-liberation. It is a voice that appears to speak the language of unfettered development, heedless of any concern for those who cannot keep up or who are subordinated as a result of the logic of underdevelopment. Now, as in the sixties, it views all talk of limits as a ruse, and it takes to the letter the half-assed promise of the dominant culture to deliver a liberated future. Distrustful of the "puritanism" of the left, and dismissive of the "techno-fear" of the "self-denying" ecofundamentalists, the New Prometheans revive the Diggers' vision of a work-free, postscarcity society, "all of it watched over by machines of loving grace."[27]

Like the illicit enclave of Chiba City's Ninsei in *Neuromancer*—"a deliberately unsupervised playground for technology itself," because "burgeoning technologies require outlaw zones" (p. 11)—the humanist counterculture of Leary et al. serves its role as an experimental sounding-board for legitimate industrial developers. As we move closer to the industry, where the cyberpunk idea has been just as enthusiastically absorbed, we find the same ecstatic language of evolution (stripped, of course, of the countercultural baggage) being spoken by legitimists of new, futuristic technologies.[28] It is conventional to speak, for example, of "generations" of technology, as if technological development rested upon natural laws of change and maturational growth rather than upon any socially available nexus of power, wealth, and expendable resources. Purely technological projections of the future are presented as if they were more infallible than social forecasting. Here, for example, is Eric Drexler, much-lauded pioneer of the "nanotechnology revolution":

> In a race toward the limits set by natural law, the finish line is predictable even if the path and the pace of the runners are not. Not human whims but unchanging laws of nature draw the line between what is physically possible and what is not—no political act, no social movement can change the law of gravity one whit. So however futuristic they may sound, sound projections of technological possibilities are quite distinct from predictions. They rest on timeless laws of nature, not on the vagaries of events.[29]

In fields like nanotechnology, with its promise of self-replicating molecular machinery, or artificial intelligence, with its fantasy of downloading the contents of the brain into robotic bodies, the humanist faith in the evolution of the species has been hijacked and entirely displaced onto technology itself.[30] As Marvin Minksy puts it, "we are sort of locked into our genetic structure. . . . there is much more potential for rapid evolution of machines than for humans."[31]

In opposition to this view of evolution, humanist critics of the project of artificial intelligence stand by their trust in the exceptional qualities of human nature, warning that the evolution of machine intelligence will be inimical to essentially human concerns and interests. Joseph Weizenbaum, author of such a critique, *Computer Reason and Human Power*, has been called a "racist" and a "carbon-based chauvinist" because of his moral preference for "the human race" over a "race of computers."[32] These only half-ironic epithets are a sample of the black wit savored by the AI community. However, if we match the AI critique with its corollary in the deep ecology movement—the charge, against humanism, of species-centrism—then we get a fuller picture of the precarious ground upon which the humanist argument stands today. The modern devotion to uniquely "human" concerns and capacities is increasingly associated with a long and destructive history in its dealings with the physical world and its nonhuman inhabitants. Its high moral ground of humanism looks less like a besieged stronghold of heroic resistance and more like a fortified position secured through aggressive domination of the natural world. A humanism that wants to police its borders with the technosphere carries with it an ugly record of policing the ecosphere. As for its global dimensions, a broader social overview of the humanist project further exposes the degree to which its historical claims have been and still are waged in the interests of white masculine power.

This sorry history everywhere undercuts the grounds of the humanist argument that the social future of technological development ought to be universally tailored to the destiny of the "human" species. Indeed, as I have suggested, this argument can easily be taken over by the technologues who preach the superior evolutionary possibilities of machines. When Drexler speaks of "the race towards the limits set by natural law" he is simply extending the old discourse of humanism to the realm of machine intelligence. This appropriation even contains its own modern version of the Faustian sin of "overreaching" that was laid at the door of early scientific humanists. In the history of modern humanism, the social costs of such a "race" have been borne by the many in the interests of a few, while scientists' ever-fluid definition of "the limits set by natural law" has been shaped and controlled by profiteers. The "cutting-edge" futures promised by evolutionary technologues will not only have to cut *someone*, they will also have to cut through *something*, and *somewhere*. From the perspective of the physical world (however you conceive it), the social footprints of humanist science are long and deep. As far as the humanist argument goes, then, the problematic which serves as the title of my next chapter—"getting the future we deserve"—is likely to be addressed only in terms of a non-secular narrative of retribution for past sins.

As a term for describing the self-conscious development of the social life of a species, humanism is increasingly corrupt—not only because of its universalist assumptions about human identity, but also because of its high-handed dealings with the physical world. The interdependence of the ecosphere and the technosphere rests upon a fully socialized set of relations that can no more be reduced to essentially humanist concerns than they can be abstracted to the quasi-divine rationality of superior technical "intelligence." It is because traditional humanism's appeal to the advancement of the species is no longer an ideal candidate for contesting the claims of technocracy that the broader critique offered by social ecology has become a radical presence in the debates about futurism. Murray Bookchin, whose name is synonymous with social ecology, has consistently argued that human domination of humans preceded, and set the model for, the humanist domination of nature associated with the modern, post-Enlightenment period. It is therefore pointless, he argues, to demonize human "society," as the deep ecologists have done, and subsequently to take the side of "nature," as if these were mutually exclusive categories of reality. More atavistic yet is the view, offered by fundamentalist critics of biocentrism, of humanity as a biological species on a par with the slug. In response, Bookchin's ecological critique is directed towards the roots of domination in specific human institutions and specific social relations—the centralized bureaucratic state, patriarchy, ageism, racism, and capitalist growth—rather than in some totalizing idea about the speciocentric domination of nature itself.[33] Bookchin reserves his faith for a fully socialized conception of humanism, drawing upon the anarchist and syndicalist history of decentralized, self-governing communities to envisage a postscarcity future in which appropriate technologies are used to realize the humanist vision of freedom.

It remains to be seen whether Bookchin's radical humanism can be reconciled with the postmodernist picture of daily contestations at the "edge" of human/machine interfaces—the "border war" that Haraway has described as a prevailing condition of our lived experience in a techno-intensive world. For Bookchin, postmodernism is simply another anti-rationalist discourse, a parallel symptom of deep ecology's anti-humanist atavism. For the postmodernist, Bookchin's humanism smacks of a preachy nostalgia for organic, even essentialist, identities, ill-equipped to engage, on the ground, with the impure realities and intensities of the new cybernetic order. There is too much to be lost politically, however, by seeing these respective positions as mutually opposed. We ought to be able to accept both critiques as dialectically linked resources for renewing the tradition of leftist futurism. In its appeal to liberationist ideals, social ecology draws upon the memory of utopian–anarchist ideas as they were practiced by historical communities, and as they still provide the inspiration for communitarian life today. The postmodernist critique of identity is a contemporary response to the social condition of modern life, teeming with the fantasies and realities of difference that characterize a multicultural, multisexual world. One is a resource of traditions and ideals persistently repressed throughout modern history; the other is a resource of tactics and mythologies rooted in contemporary experience. Both ought to be seen as renewable resources, dialectically sustaining each other whenever we need to call upon them.

CHAPTER FIVE

GETTING THE FUTURE WE DESERVE

Few people can doubt that the "future," as we know it, has changed in the last year or two. Two sets of images, in particular, encapsulated this transition. The liberated populations who gathered in city squares throughout Eastern Europe offered the spectacle of a new kind of revolution, neither propelled by a faith in the historical certainty of its future nor resisted by a faith in the past on the part of the old ruling guard. The second set of pictures portrayed a physically dying planet, exhibited in various grisly poses of ecological depletion, and circulated by all sectors of the image industry, often in spots and contexts reserved for exploitation atrocity fare. Different as these images were, each told a fundamentally similar story about the exhaustion of the narratives of progressive futurism which have occupied a central place in Western culture for the last two centuries. In the latter half of that period, the socialist project of creating a better future had been entrusted to a widely held faith in historical "laws," administered by technological growth and development, and the centralized, rational planning of society. "Social progress" could only come about in a future that would be a true progression, the result of a rational use of technical resources unfettered by the irrational needs of capitalist growth and accumulation.

For the left, the lessons to be learned from these images speak acutely to our traditional responsibility to think about a better future. Our visions of future "freedom" will have to be fiercely conscious of limits. As the final collapse of plan-oriented state socialism and the spectacle of ecological depletion make clear, our commitments to social growth must everywhere be mediated by limits to technological growth, to human use of the physical world, and to the "scientific" planning of political and civil life. In short, we must abandon the old idea that the politics of social growth has at its command unbounded resources, both human and natural, that can be marshaled in the service of universal and univocal ideals.

In this chapter, I will be discussing some of the circumstances under which intellectuals and activists are faced with the task of renewing left futurism today. Without a set of ethics about the shape of possible futures, the left threatens to lose whatever coherent shape it still commands in the present. In the absence of any invitation to participate in the future, people tend to respond to appeals to their short-term interests. The arguments presented here presume a radical break with the certainties of one classic version of left futurism—the tradition of Marxist historical teleology, with its roots in the Enlightenment faith in scientific progress through technical mastery of the natural world's resources. Indeed, my arguments resonate more with the traditional aim of utopian socialism, that of finding a credible language and imagery to represent the idea of a more radically democratic future; a horizon of expectations for *different* people to live by and act upon, with some measure and promise of real gratification. The term "different" here encompasses not only the old utopian inspiration that people can live in radically different ways under changed social circumstances but also the significance of the new politics of difference as it relates to people defined by differences of gender, race, class, nationality, and sexual preference.

In the pages that follow I will tell a number of related stories about the history and current conditions of futurist thought, which, for over a century, had been considered the natural province of Marxism and socialism. One of these stories concerns the postwar "bourgeois" appropriation of traditionally left futurism in the bureaucratic form of *futurology*, a social science of systems analysis created to facilitate military and industrial planning and fully institutionalized today as an instrument for acquiring strategic military or corporate advantage. Postwar futurology took a number of forms, but I will be particularly concerned with the liberal versions that developed in the 1960s and 1970s, which proposed a concept of "possible" futures that helped to explain—and legitimate—many of the "flexible" planning features of a new economic order that is increasingly referred to as post-Fordist. Another story concerns the career of environmental futurology, initiated by ecologists and social forecasters in the early seventies as a series of statistically based surveys of the deterioration of the world's resources. This kind of futurology was initially an alternative to the growth-driven logic of military and industrial planning, and while it has been largely absorbed into the corporate project of global management, it retains its use as a genuine resource for contesting the capitalist logic of ceaseless growth and accumulation. My third story is about the renewal of utopian futurism in the new social movements, the natural home today of liberatory passions and visions of the future that will respect differences among people. In the preceding chapter, I have already dealt with a fourth story, in recounting some of the dystopian responses in popular culture and consciousness to these changes in the postwar landscape of futurism, particularly the shift from Cold War scenarios of swift nuclear annihilation to the more recent, dark eco-futures predicated upon slow, environmental deterioration and collapse.

The need to recount these stories stems, first of all, from a major shift in left thought about the principles of Marxist futurism, once thought to be determined by natural laws of history. There is no longer any single Marxist story about social growth governed by a "science" of social progress that would deliver a better world; no futures are inevitable, not even those that appear, statistically, to be just around the corner. Moreover, the rationalist planning of left technocratic radicalism associated with "progressive futurism" had less and less to offer in the way of *diverse* gratifications. Its rigid blueprint for a "better world" was not flexible enough to accommodate the desires shaped by the full range of social, sexual, racial and ecological differences, while its almost religious belief in the top-down engineering of social life took a heavy toll on available resources, again both human and natural. Finally, the imagery of "mass" life and "mass" culture so established in the repertoire of centralized state paternalism (whether it was celebrated or vilified, by the left or by the right) masked a cruel disregard for the popular hunger for democratic participation and self-determination. Unlike the older rationalist models, new futures with progressive credentials carry no guarantees about predetermined goals and shapes. On the one hand, there is something intimidating about such open-ended futures without an architectural blueprint—the "Marxism without guarantees" of which Stuart Hall often speaks; Marxism without "the masses," without any single progressive bloc to rally around, or universal value-systems to invoke as a reliable guide. On the other hand, there is nothing inherently disabling about these requirements; they speak more to the progressive renewal of socialist traditions than does the guilt-ridden left custom of confessing to historical errors. Nor does their appeal lie in starting *ex nihilo*—the fateful impulse of historical avant gardism. In fact, the appeal is primarily to *continuity* with the work, thought, and activism of the new social movements that have become the radical political lifeblood of our times, each with its own distinctive utopian traditions defined by the histories of gender, ethnicity, sexual preference and biological diversity.

Aside from these developments in socialist theory and practice, the postwar history of elite futurology, combined with the effect on the populace of the corporate state's manipulation of the "nuclear threat," has further eroded the left's once solid claim on the future. The emergence of the ecology movement in the late 1960s offered a chance of reasserting and reimagining the left's stake in futurism. By exploring the ways the future has been alternately colonized by elites and contested by progressives in the two decades since then, this chapter will try to assess the opportunities that exist today for a renewed left futurism that can use its resources in a more sustainable way, not only with respect to the natural world but also by responding to popular desires for a more creative and less standardized way of life.

"BOURGEOIS" FUTUROLOGY

Attempts today to provide a progressive imagery for the future will find the terrain tough going. Over the last forty years, the "future" has been heavily populated by traditionally anti-progressive interests. It has become the natural habitat of technocratic elites; a lucrative haven for financial speculators; an indispensable tool in the politics of crisis management; a professional training ground for militarists; the next frontier for free-marketeers; and the locus for "thinking about the unthinkable," to use Herman Kahn's notorious phrase for describing the logistics of post-nuclear survivalism. Hitherto considered the undisputed home of left-wing utopian or "scientific" socialist thought, corporations and the military establishment have come to devote enormous energies to the future, setting large numbers of futurologists to work in academia and in foundations, institutes, and think-tanks established to provide legitimation for the policies of the modern corporate state through the use of the new intellectual tools of systems analysis, operations research, information technology, and simulation modeling.

In the United States, futurology originated in the special relationship established between military weapons planners and civilian scientific advisors in "operations research" during the Second World War. Anxious to further exploit these military–civilian ties in the study of V-1 and V-2 rocket technology and intercontinental air warfare, General H.H. Arnold, the Army Air Force Chief, used leftover funding from his war budget to establish Air Force Project RAND (Research ANd Development) at the Douglas Aircraft Company in 1946. Two years later, with additional aid from the Ford Foundation, RAND was incorporated as an independent institute for futures research. RAND maintained its client ties to the Air Force Project, researching the techniques of military and political warfare, and producing studies for the Air Force's Long-Range Technological Forecasts, which defined the needs of strategic military advantage for military elites and lawmakers in the business of procuring public money for weapons development. The RAND Corporation thus set the model for subsequent outfits like the Institute for the Future, the Hudson Institute, Systems Development Corporation, and the Institute of Defense Analysis, think-tanks in the business of political advocacy, guided by the permanent war economy's frame of reference. Kahn's work at the Hudson Institute (founded with the motto, "National Security—International Order") produced Cold War classics like *On Escalation: Metaphors and Scenarios* and *Thinking about the Unthinkable* which were leading examples of the militarization of "future scenarios." The story of intellectuals in these institutes is that of technocratic elite formation in the US, first envisaged in the more progressive moment of the New Deal (Roosevelt's Brains Trust), and in the radical technocracy movement described in chapter three. Research studies produced for public clients like government agencies continued to bear the mark of their origin in military operations research since they appealed to a process of elite decision-making whose automatic command structure was aimed at executing planned objectives with maximum efficiency.

In the elite milieu of these institutes and think-tanks, there was no role for popular or democratic decision-making about the future. Futurology was pursued as a technocratic resource for military–industrial elites and political policymakers, legitimating massive military spending by promoting the idea of winnable nuclear wars or imagining survivability in life-after-death nuclear scenarios. Its methods of information analysis, originally applied to weapons forecasting, were increasingly adopted for social forecasting by government agencies and corporations in the business of designing the future of whole social systems. The history of professional US futurology cannot, however, be understood simply as a conspiracy among corporate–military elites and their hireling intellectual workers. It is also an important chapter in the story of postwar liberalism. In the 1960s, a significant number of liberal, publicly minded intellectuals entered the field of futurology on the back of a growing wave of anti-technocratic sentiment among dissenting sectors of the population. With the intent of broadening consensus opinion about the shape of the future, the new kinds of think-tanks included a larger share of experts—sociologists, anthropologists, political scientists, social psychologists, educators, philosophers, and even poets—who were less directly tied to military and corporate interests, and who were called upon to address social and ethical concerns in the public interest. International nonprofit organizations like the World Future Society (USA), IRADES (Italy), the Committee for the Next Thirty Years (UK), *Club de Amigos de la Futurologia* (Spain), and *Association Internationale des Futuribles* (France), *Institut Für Zukunftsfragen* (Austria), and the Club of Rome (Italy) were established; public organizations like the Commission on the Year 2000, the Committee for the Future, and Mankind 2000 were set up; and journals like *The Futurist* (USA), *Futures* (UK), *Futuriblerne* (Denmark), *Analysen und Prognosen* (Germany), *Analyses & Prévisions* (France) and *Futuribile* (Italy) flourished. Private companies like Forecasting International, Inc., Futures Group, Data Resources, Inc., university foundations like the Stanford Research Institute, the World Resources Inventory, and the Institute for Alternative Futures, and ecologically minded public interest groups like the Worldwatch Institute, Resources for the Future, and others followed suit. Eventually, in 1975, a

Futures Research Group was set up in Congress in the wake of the establishment of the Office of Technology Assessment, an environmentally minded successor to the tradition of technological forecasting. Among these groups, the intellectual cult of predicting the year 2000, in particular, took on all of the elements, as Daniel Bell put it, of a "hoola-hoop craze."[1]

Although many of these groups were instituted in the public interest, and with the support of public monies or Ford or Carnegie Foundation grants, some were directly tied to and paid for by business interests, increasingly reliant upon the advisory support of academic experts. The explosion of forecasting activities on the part of these groups testified, on the one hand, to the professionalization of the long gentleman-amateur tradition of intellectual prophecy, a tradition pursued historically by isolated predictive thinkers like Francis Bacon, H.G. Wells, Jules Verne, and George Orwell, continued by high-profile scientists and engineers in the (turn-of-the-century) golden age of invention; and preserved more recently in the prognosticatory subculture of science fiction literature. On the other hand, the futurology boom also marked the consolidation, extension, and ultimately, supersession of the scientific management principles that prevailed during the age of industrial Fordism. From the early 1920s to the early 1970s, the economy of the United States and other advanced capitalist countries was largely organized around so-called Fordist principles of production, initially developed by Henry Ford. Central to these principles were a set of scientific and rationalist methodologies—rationalized mass-production and consumption, scientific management of time, inside and outside the workplace, and the social management of class relations between capital and labor—that had their origin in nineteenth-century philosophies of the "natural laws" of concrete social processes. Marx and Engels had adopted such scientific explanations of social processes to argue against the utopian socialists who called for qualitative social change in the present as well as in the future. Marx, Engels, and their followers believed that "scientific" knowledge of the objective laws of social change and economic production was finally a sufficient cause for knowing the future; no amount of critical and activist attention to present-day change would *significantly* alter that future. In the Marxist tradition of scientific socialism, the present was seen as a phase of transition to a determinate future: the final collapse of capitalism and its inevitable replacement first with socialism and then with communism.

Far from urging on the collapse of the present system at some inevitable crisis point in the future, the goal of futurology had been to manage and control the future through the use of systems analysis, computer databases, and modeling developed earlier by military–industrial forecasters. Futurology, then, shared certain structural affinities with the Marxist belief in natural laws: both interpreted the future as a necessary, causal element of the present. In their models and scenarios, many of the hard science forecasters tended to anticipate a continuation of current trends, as if they constituted a "natural" basis for extrapolating a picture of likely social futures. Kahn and Anthony Wiener's *The Year 2000: A Framework for Speculation for the Next Thirty-Three Years* (1967) was the exemplary survey of trend extrapolation, based on past data and the assumption that no major changes in social, physical, economic and political configurations would occur in the future. Consequently, the facts, statistics, and indices of production, employment, and population that were selected for such models and surveys tended to reflect the existing structures of the liberal market economy of corporate capitalism. Long-term trends reflected the increasing concentration of wealth, political power, scientific knowledge, and military capability along the lines of the existing system. Such "surprise-free" scenarios, pictured from the elite standpoint of those with the greatest investment in the continuation of such an economy, were incapable of assessing the impact of future social changes that radically diverged from the basic framework of liberal capitalist society, let alone incorporating the need for, or possibility of, a different kind of economy. The result was to naturalize further the existing features of a laissez faire system. As Bell, again, put it so warmly, in his preamble to the collected symposia discussions of the Commission on the Year 2000 (set up by the American Academy of Arts and Sciences), hope for the future "resides, first, in the marvelous productive capacity of our system to generate sufficient economic resources for meeting most of the country's social and economic needs."[2]

Besides providing a "scientific" demonstration of the shape of a liberal capitalist future, sixties futurologists offered a new knowledge commodity: the opportunity to "explore" alternative futures within the confines of the existing system. The new feedback variable in the future was supposed to be futurology itself. Predictions would have an actual impact upon the future because, in assessing and measuring the pattern of future events, they would have a dynamic influence on social and economic decision-making. Futurology, then, promised a more interventionist role in the business of managing and controlling possible or alternative futures that were more difficult to envisage from the standpoint of the present. Futurology groups researched the shape of such futures and presented them in the form of possible options. For example, an oil company might be presented with a number of future scenarios that combined traditional energy forecasts with the likely influence of a host of new public and legal concerns about environmental issues; the scenarios would also account for various alternative social, political, economic and technological environments in the future. Whether or not this kind of research was undertaken directly for a corporate client, it is easy to see how it serviced the corporate need for contingency planning. Such models reduced corporate anxieties by rationalizing the scope of possible control over future trends. The goal was intelligent management and control of the unpredictable. The World Future Society suggested spheres of influence to map out the geography of the future: "immediate futures" (up to one year) were uncontrollable because they were dictated by the past; "near-term futures" (one to five years) could be partially controlled by present action; "middle-range futures" (five to twenty years) could be almost completely controlled and chosen from the present; "long-range futures" (twenty to fifty years) could be "seeded" from the present; while "far futures" (fifty years and beyond) were largely invisible and thus uncontrollable.[3] In the process, the futurologists developed new techniques of prediction. A consensus technique, called DELPHI, was developed by Olaf Helmer and Norman Dalkey at RAND for group forecasting that involved interactive opinion-making among a group of experts. Although many of the new techniques for forecasting preferable futures were causal models, they were more dynamic than the trend and growth curve models traditionally used for extrapolation. Cross-impact analysis, developed by Ted Gordon at the Futures Group, was designed to gauge the effect of new factors upon existing ones. Other tools included decision trees (graphically depicting various branches leading to possible futures), scenarios, and simulation modeling of complex systems. Taken together, these methods were aimed at theoretically anticipating all possible chains and patterns of events, a goal that went beyond assessing *probable* futures to one of preparing *possible* and *preferable* futures. Because they were undertaken with the aim of producing knowledge that would influence the shape of the futures, most of these studies were written up to appeal to policymakers and corporations as an information resource that would help the political and corporate elites to plan and create preferable futures. One royal question, of course, was left begging: whose preferences were being served?

At the prodding of progressive groups, some of the more liberal groups included humanistic, value-oriented futurists in their discussions. Discussion of values bridged the gap between American futurology hitherto dominated by "hard data," and the broader European futurist tradition of human forecasting and philosophical speculation pioneered by figures like Fred Polak, Ossip Flechtheim, Bertrand de Jouvenel, Robert Jungk, and Eric Jantsch. In addition to these European social democrats, North American liberals who were regular participants in the new international forums included the likes of Kenneth Boulding, John McHale, Margaret Mead, Isaac Asimov, Marshall McLuhan, Arthur C. Clarke, Willis Harman, and Hazel Henderson. Their common task was to generate a debate about the values and ideological assumptions around which desirable futures could be planned. While some of the Europeans like Jouvenel clung to Saint-Simon's vision of a society entirely and rationally guided by expertise, most rejected the technocratic model of elite decision-making and prediction and made the case for greater public participation. Some, like Robert Jungk (long-suffering anti-fascist and anti-nuclear activist from Germany) and Johan Galtung (from the International Peace Research Institute in Oslo), were sympathetic to the new youth movements and agreed with Arthur Waskow (from the Institute for Policy Studies in Washington) that technocratic "planning was clearly a way of helping those who now hold power to know what they must do in order to keep holding power thirty or fifty years hence. What must they change, where should they beat a strategic retreat, what new organizations and technologies should they invent, when can they hold the line?"[4] Waskow, a fierce proponent of the demands of the student movements and countercultures for a radically different future, argued for experimental, action-oriented community attempts to build "chunks of the future" in the present, from the bottom up, without the permission of elites. The model for these experimental futures was already flourishing in the utopian countercultures, based upon radically alternative social values and lifestyles, and organized by "experts" schooled in "creative disorder" rather than in technological forecasting.

Waskow's call for direct action toward participatory futurism went beyond the attempts of the value-oriented liberals to fine-tune elite futurology by "including the public" in futurist decision-making. Liberal anguish about the anti-technocratic challenges made by the late 1960s movement for participatory democracy fed directly into the elite tradition of crisis-management by experts, accustomed to coopt all democratic challenges to its mechanisms. In the proceedings of the Commission on the Year 2000 (which included, among others, Daniel Bell, Eugene Rostow, Zbigniew Brzezinski, Erik Erikson, Samuel Huntingdon, Herman Kahn, Daniel Moynihan, Ithiel de Sola Pool, David Riesman, Wassily Leontief, and Margaret Mead), for example, elite concession to these challenges seldom got beyond the suggestion of including other kinds of experts—more "imagination people" to balance the "hard data people"—in the dialogue about the future. For the most part, the inclusion of more visionary, liberal social critics only created the appearance of a broader public dialogue in futurist forums like that of the Commission.

The bestselling futurologist Alvin Toffler was the most successful and popular of these liberal critics. His arguments tapped into the heady glamor of the futurist moment and simultaneously appealed to the liberal and countercultural rejection of top-down technocratic planning. In a series of enormously popular books published from the late sixties onwards, including *Future Shock* and *The Third Wave*, Toffler produced the most systematic attempt at describing how the anxieties of a mass population about the future might be addressed and managed in ways that seemed more humane than the traditions of technocratic planning as practiced in the socialist command economies or among the corporate capitalist elites.

Toffler earned the respect of "responsible" liberals through his harsh critique of the "anarchic" planlessness espoused by utopian communes of the late 1960s. The utopianism of these communes, Toffler argued, was guided by preindustrial, pre-technocratic values and thus offered little in the way of a concrete vision of a better future. Toffler set up shop in the counterculture's planless vacuum, issuing calls for an "anticipatory democracy" that would be able to absorb the most radical effects of a society changing too fast to avoid what he called "future shock." Toffler's work had a multifaceted appeal: his ageist attack on the remoteness of the "elderly technocrats" appealed to generational rebels; the suggestions of his Committee for Anticipatory Democracy (instituted in 1975, and active in state and urban Anticipatory Democracy groups all through the Carter administration) for public plebiscites and "town hall meetings" to discuss the future appealed to liberal communitarians; and his call for humanized, flexible planning appealed to the corporate liberals, eager to transcend what they saw as the more rigid, obstructive features of Fordism such as centralized planning, or the social contract between capital and labor.

Toffler's watchwords, like "future shock" and "anticipatory democracy," spoke to the anxieties, fears, and desires of people who saw planning and decision-making as remote, hierarchical, and undemocratic processes. But these ideas were arguably most successful in unwittingly providing a popular ideological vehicle for the vast transformations ushered in by corporate capital's economic, social, and cultural restructuring that led to the breakup of the Fordist social contract between capital and labor. The kind of arguments made by Toffler and other liberal futurologists about "open" and "flexible" futures have to be seen, in retrospect, as an *explanatory rhetoric* for the new economic and cultural arrangements that have come to be known as post-Fordist. With the dissolution of the postwar Keynesian state's compromise, in which large-scale planning had benefited capital and labor alike, the new emphasis on flexibility in labor processes, marketing, and technological development transformed the future into a fluid set of options for planners to choose from. By contrast, the contours of the future had been settled, if not always guaranteed, by the Fordist promise of long and slow growth. Futurologists who talked about "preferable" and "flexible" futures were speaking a language that was siren-sweet to the ears of capital, so long used to hearing the prosaic liturgy of the social contract, the social wage, and social regulation. Professional futurists, even those who claimed to represent popular desires for a less standardized way of life, thus contributed in no small way to launching the volatile post-Fordist future of flexible accumulation.

Consequently, the leading edge of the post-Fordist order was to become a high-risk game of venture capital, which guaranteed losers nothing more than a "safety net" if they did not prove agile enough to keep up, and where "flexibility," in the current climate, is often simply a code-word for deregulation, capital mobility, deskilling, union rollback, the feminization of poverty, and overt class war from above. If left futurism today succeeds in renewing itself in such a climate, it will not be out of any second-string obligation to follow the "creative–destructive" lead of the post-Fordist cutting edge, but rather out of the belief that the diversified post-Fordist world of production and consumption was not an autonomous capitalist creation, and that many of the features of that world are deeply rooted in, and took their inspiration from, popular desires for a more diverse and less standardized way of life.

FUTURES TRADING

Alongside these complex developments in social forecasting and future-management lay another area of economic life in which attempts to colonize the future profitably had gained ground: the futures market in speculative trading. Futures exchanges today are financial markets in which commercial firms, hedgers, and speculators shift part of the risk of price change to speculators who assume the risk for the chance to earn a profit on their venture capital. Lewis Mumford has pointed out that speculation in "futures" was already well developed by the middle of the sixteenth century in the international money economy; "non-commodities, imaginary futures, hypothetical gains" were an established source of short-term profit, and were an integral part of the new "tyranny of time" around which mercantile capitalism reorganized everyday commercial culture.[5]

It was not until after the Civil War in the US, however, that the futures market was formally organized in commodity trading. Future contracts, which depended upon some future delivery of a commodity, were accepted as a "hedge," a form of insurance against the price of a commodity changing too much before it was delivered. Speculators argued that these "fictitious sales" were an important way of regulating the volatility and instability of a free market, and therefore constituted a natural, evolutionary development in the management of a complex economy: "Speculation," said Justice Wendell Holmes, "is the self-adjustment of society to the probable. Its value is well-known as a means of avoiding or mitigating catastrophes, equalizing prices and providing for periods of want."[6] Populist and agrarian interests, particularly in the farm belt, were marshalled against the new power of the speculators, whose activities were remote from the daily economic life of farm commodity producers. Over the next fifty years, the gradual erosion of the populist political challenge, laced with the strong moralistic appeal of anti-gambling sentiment, was concomitant with the supplanting of agrarian interests and power in Congress and in national life generally. As the futures market moved from the commodity pits to stock markets, agrarianist critics who had given voice to the widespread populist ambivalence about futures speculation as a form of economic parasitism, responsible for increasing rather than reducing risks, found themselves ever more removed from the actual location of futures activity.

Consequently, self-regulation took the place of the organized political pressure that these critics had brought to bear upon the market. Eventually, the cash commodity began to supplant trade in pork bellies and the like as a trading standard. Between 1970 and 1985, the volume in financial futures contracts increased over tenfold, and by 1985, financial futures and trading markets accounted for over 60 per cent of total trading volume. After the oil crisis of 1974, futures markets became an acceptable form of risk management outside of agriculture. The distinction between securities and futures markets—commodities and stocks—began to dissolve as futures contracts flourished in almost every area of investment and trading. When the obligation to make real cash deliveries is withdrawn, then there is little difference between shifting risk and shifting capital.

Today, the futures game is played out in almost all areas of economic life, and has come to take a central role in the economic cost–benefit game of public policy. In the Reagan years, it was accomplice to the spectacular, high-risk developments of junkbonding, asset-stripping, and corporate raiding, and its role in the global finance economy today affects the economic life of all nation-states dictated by the vicious demands of foreign and international debt. Not even its most diehard advocates can claim any longer that the futures game is a sane device for regulating the risks inherent in an unstable market economy, or that it is a useful feedback mechanism in the business of politically controlling such an economy.

Given this current picture of over-speculation on the global markets, we cannot avoid seeing how the "future" has already been bought up and sold many times over, at a level that will largely determine how hopes of alternative economic futures can be built. In the US, public consciousness about the massive yearly interests paid on the national debt and the consequences of the Savings & Loan debacle has reinforced the perception that the present will be in hock to the future for generations to come.

ENVIRONMENTAL FUTUROLOGY

For the left, the postwar story of futurology that I have briefly recounted has been somewhat different. For "scientific" socialists, postwar futurology is simply an attempt to "steal" the future from the left. Consider Georgi Shakhnazarov's critique of bourgeois futurology as "non-scientific" in comparison to communism's institutional futurism:

> Futurology originated in the West as a specific academic discipline to provide an immediate alternative to the Marxist–Leninist, communist ideas about the future society. It was no natural offspring born at a point where a new branch of science hives off from the general body of knowledge. It was rather a peculiar hybrid artificially produced by the bourgeois social sciences (political economy, philosophy, sociology, law, political science, and psychology) to meet a generously rewarded demand. Unlike any genuine science that arrives at the truth at the end of the road, futurology has an anteriorly posited result to which it must suit the facts.[7]

In intellectual circles touched by the utopian tradition, with less faith in the scientific certainties of historical materialism, the impact of radical skepticism among Western Marxists had taken a heavy toll. In *The Image of the Future* (1961), his elegant history of futures past, Fred Polak complained about the "degeneration of the future" in the postwar period. For the first time in Western history, he argued, progressive advocates of social and cultural consciousness lacked a positive image of the future, not only in time, as in the tradition of Enlightenment progress, and in space, as in the utopian tradition, but also in spiritual matters, where the much older eschatological traditions had run out of steam.[8] For Polak, the nihilistic cult of existentialist philosophy—"a future without a future"—was an especially trenchant symptom of intellectual malaise. He may have come to hold similar opinions about the structuralist and poststructuralist successors of existentialism, since there was arguably less that was recognizably utopian in their deconstruction of the limits of human subjectivity and agency. Polak, like Robert Jungk and other socialistically minded members of the first European postwar generation of professional futurologists, called for a creative minority of intellectuals to "seed" the future with more idealistic imagery as a way of repairing the damage done by the counter-utopianism of the day.

If Polak's concerns still ring true for the left today, then his recommendations will have to be pursued in a changed historical context. First, the concept that a "creative minority of intellectuals" should play such a key role is increasingly perceived, at least on the left, as elitist and vanguardist as regards the claim to speak for society as a whole. Second, the growth of the ecology movement has broadened interest in the future by portraying a future in which everyone could be seen to have a participatory stake. Among the new social movements that emerged in the late 1960s and early 1970s, the ecology movement was the one most tied to an explicit set of theses about the future: how best to avoid a disastrous, and generate a better, future. The environmental critiques of macro-technological development, hard energy dependency, the exploitation of nonrenewable resources, and of political structures like the centralized nation-state and the permanent war economy all harbored corrective visions of a more ecotopian future. Some of these visions were fully fleshed out in fiction—for example, in Ernest Callenbach's novel *Ecotopia* (1977)—while others were put to the test in the many experimental communes, set up to foreshadow a decentralist, self-sufficient future in a steady-state world, that have continued to flourish within the New Age movement. Thinking and acting ecologically were futurist in a way that politics had not been since the heyday of the mass socialist movements in the earlier part of the century.

Because futurism was such an inherent feature of environmental politics, and because the ecology movement was virtually unique in relying upon appeals to science for proof of the justice of its claims, environmental futurology was rapidly established as a central activity of alternative futurist groups like the Rocky Mountain Institute, New Alchemy Institute, TRANET, World Resources Institute, Farallones Institute, Planet Drum Foundation, Worldwatch Institute, Woods Hole Research Center, and in the work of ecology movement intellectuals like Paul Ehrlich, Hazel Henderson, Garrett Hardin, and Lester Brown. While institutional futurology had earlier incorporated ecological concerns, most notably in the work of John McHale at Buckminster Fuller's World Resources Inventory in Carbondale, the ecology movement had a swift impact on corporate futurology, where future modeling and scenario projection has been an environmental genre in its own right ever since the Club of Rome's famous 1972 global report, *Limits to Growth*. A highly exclusive, international association of industrialists, scientists, economists, educators and statesmen brought together by the Italian economist Aurelio Peccei in 1968 to redirect the "mismanagement" of the current world system, the Club of Rome made use of the latest computer simulation techniques at MIT—Jay Forrester's system dynamics—to predict world trends in five major areas: population, industrialization, food supply, depletion of nonrenewable resources, and pollution. *Limits to Growth*, the book based on this study, tried to demonstrate on a global scale the finite boundaries to the expected exponential growth of these factors. Predicting an uncontrollable decline in economic, ecological and cultural conditions if exponential growth were not sacrificed in favor of sustainable global equilibrium, the report, much criticized and rebutted by establishment futurologists for its "pessimistic" conclusions, outlined the many problems for which technological solutions do not and cannot exist, recommending a "Copernican revolution of the mind" to address the near horizon of limits to growth and wasteful lifestyles.[9]

The general effect on public consciousness of the new eco-futurology was to emphasize the deterioration rather than the annihilation of the future. Consciousness about the "recession" of the future qualitatively shifted when the prospect of swift nuclear annihilation was replaced by the relatively slower and more painful prospect of ecological collapse; a displacement reinforced by the analysis of a "nuclear winter," the environmental forecast for a post-nuclear world produced by a team led by Carl

Sagan and Richard Turco.[10] More important, the political effect of such reports and analyses showed that an entire "way of life" was now in question. The whole system of maximized production, ceaseless growth, and reckless consumption associated with this (Western) way of life was something for which everyone identified with this lifestyle could, in principle, feel responsible.

One important official US government reponse to the new "slow" catastrophic future took the form of a futurological report on the global condition that President Carter commissioned in his Environmental Message of May 1977, after a whole decade of environmental activism, inaugurated in the mass mobilization of Earth Day 1970, had inspired official measures such as the creation of an Environmental Protection Agency, the banning of pesticides, and the funding of alternative energy research. Intended to serve as "the foundation of our long-term planning," the *Global 2000 Report* aimed to collate the results of modeling from a whole spectrum of different government agencies and to gauge the "continuous influence that all the elements—population, resources, economic activity, environment—have upon each other."[11] Basing its picture of the year 2000 on conservative (that is, optimistic) estimates, the *Report* included predictions about the starvation of over a billion people, the disappearance of half the world's forests and one-third of its arable land, a 20 per cent increase in soil erosion and desertification, the possible extinction of up to 20 per cent of all plant and animal species, a 35 per cent decline in water supply, and a chronic increase in salination of land and water.

The official congressional subcommittee that responded to the Report summed up its findings as follows:

> If present trends continue, the world in 2000 will be more crowded, more polluted, less stable ecologically and more vulnerable to disruption than the world we live in now. Serious stresses involving population resources and environment are clearly visible ahead. Despite greater material output, the world's people will be poorer in many ways than they are today. For hundreds of millions of the desperately poor, the outlook for food and other necessities of life will be no better. For many it will be worse. Barring revolutionary advances in technology, life for most people on earth will be

> more precarious in 2000 than it is now—unless the nations of the world act decisively to alter current trends.[12]

Though it paid lip-service to the need for global cooperation to "avert the catastrophe," including support from the private sector and multinationals, the response of the subcommittee formed to "interpret" the Report was loaded with nationalistic preoccupations and assumptions, reminding its readership that impoverished Third World countries "do not make good customers" for US trade, and that the further ecological impoverishment of these countries threatens "our own security and standard of living." Ironically enough, these observations provided subcommittee members with an opportunity to advocate more free trade—"a more liberal trading order"—ostensibly to benefit all countries involved. In a classic instance of blaming the victim, the subcommittee document went on to suggest that most ecological damage is done by "desperate peoples" in Third World countries—a comment that hardly bears justification if one considers that the US, with 6 per cent of the world's population, consumes as much as 40 per cent of the world's nonrenewable energy and resources.

Despite its willful misconstrual at the hands of congressional crisis-managers, the *Global 2000 Report* itself was sufficiently critical of the outcome of "present growth trends" that incoming President Reagan employed two right-wing futurologists, Herman Kahn and Julian Simon, to rebut its findings in a report entitled *Global 2000 Revised*. Reagan's response heralded a disastrous decade of inertia and cutbacks on the environmental front. The decade also saw the reascendancy of technocratic decision-making aimed at privatizing and militarizing scientific and technological research in the name of international competitiveness. By reaffirming the corporate state, harnessing public institutions to private ends, and abdicating the obligation of government to act in the public interest, the Reagan administration established the principle of corporate access to scientific knowledge as the most important factor in shaping the future.[13] Concurrent with Reagan's response to the *Global 2000 Report*, the Department of Defense initiated a vast future modeling project to aid the Joint Chiefs of Staff in long-term planning of future weapons systems. In doing so, the military elites were reasserting their claim to modern

American futurology, a claim staked out in the late 1940s, and temporarily undermined by the enterprise of liberal futurology in the 1960s and 1970s.

Despite the massive setbacks of the Reagan years, the impact of environmental futurology has gone from strength to strength. Today, it exerts a considerable influence over planning and decision-making at the highest levels of the new global politics. For example, the wide-ranging report of the UN-sponsored Bruntland Commission, *Our Common Future* (1987), with its prospectus for "sustainable development," has generated changes in government policies throughout the world.[14] On the other hand, the powerful influence of such reports is doubtless being absorbed and exploited by the politics of "planetary management" practiced by the World Bank, the International Development Association, and transnational corporate cartels bidding for the role of planetary protector by pledging a global conscience in the name of eco-capitalism. Ecology, at this level, becomes one more hand in the high-stakes game increasingly played by the new global managers, operating beyond the regulatory control of sovereign nation-states.

A major accomplishment of environmental futurology has been to focus public attention on the international dimension of the ecological crisis. This awareness, best encapsulated in the cluster of fears and anxieties evoked by the term "global warming," signals that globalism, as an everyday idea, has finally broken the surface of popular political consciousness, so long bounded by short-term interests and nationalistic anxieties or desires. The ecology movement has forged a decisive link between individual actions and responsibilities and global or planetary concerns.

Social ecologists balk at the idea of "sustainable development" with good reason, arguing, as Murray Bookchin has put it, that "capitalism can no more be 'persuaded' to limit growth than a human being can be 'persuaded' to stop breathing."[15] But there are few good reasons for not using the prominence of environmental reports such as *Our Common Future* as a helpful resource for urging ecologically minded claims at all levels. They may represent the elite product of a politics of information and knowledge, and they may be cited as further examples of the "colonization" of the ecology movement by its corporate adversaries, but the cogent impact of this kind of environmental futurology has already forced more governments and corporations to modify their policies and practices than any other political or social movement in such a short space of time. These successes are the result of persuasive appeals to scientific evidence; they are not a byproduct of the corporate "benevolence" that has become such a pervasive feature of sophisticated eco-marketing. But dependence on appeals to scientific information and knowledge alone does not make a social movement, especially one with a widespread popular appeal. The experience of the ecology movement is that environmental futurology is easily cooptable if it does not take its lead from public and popular activism that "does" science by practicing it on the ground and in the daily life of local communities. For the seeds of a more ecotopian future lie as much in people's everyday idea of the future they would like to inhabit as in the complex maps drawn by computer modelers.

THE PRESENT FUTURE

So where does this recent history of planning and imagining the future leave us? The task of renewing left futurism today seems to be burdened with the twin legacy of a large-scale colonization of the future by elites and a powerful, dystopian impulse on the part of progressives to abdicate thinking about the future.

In his book *The Year 2000* (1983), one of the best and most comprehensive of recent socialist writings on the future, Raymond Williams described today's elite futurological politics, with its militarized rhetoric of strategic advantage, as "Plan X":

> [I]t is different from other kinds of planning, and from all other important ways of thinking about the future, in that its objective is indeed "X": a willed and deliberate unknown, in which the only defining factor is advantage. . . . Thus while as a matter of public relations [the Plan X people] still talk of solutions, or of possible stabilities, their real politics and planning are not centred on these, but on an acceptance of the indefinite continuation of extreme crisis and extreme danger. Within this harsh perspective, all their plans are for phased advantage, an effective even if temporary edge, which will always keep them at least one step ahead in what is called, accurately enough, the game plan. . . . At the levels at which Plan X is already being played, in nuclear-arms strategy, in high-capital advanced technologies (and especially information technologies), in world-

> market investment policies, and in anti-union strategies, the mere habits of struggling and competing individuals and families, the mere entertainment of ordinary gambling, the simplicities of local and national loyalties (which Plan X, at some of its levels, is bound to override wherever rationally necessary) are in quite another world. Plan X, that is to say, is by its nature not for everybody. It is the emerging rationality of self-conscious elites, taking its origin from the urgent experiences of crisis-management but deliberately lifting its attention from what is often that mere hand-to-mouth behaviour.[16]

As a riposte to the rationality of Plan X that imagines the "natural landscape" of the future as a territorial war of position, Williams argued that alternative agendas for the future, if they are to be serious, have to be "*more* rational and *better* informed," and "morally much stronger." The stockpiles of data, persuasive information, and scientific knowledge amassed by the ecology, peace, and feminist movements are exemplary, in his view, of the rational resources needed to construct an alternative futurism.

But Williams can also be read as pledging yet another round of trust to the rationality of elite expertise—in this case, the expertise of elites *within* the oppositional movements—even though he does so in good faith as a means of projecting alternative futures. This argument only goes so far. Although the vanguard tradition of the social expert, architect, or planner who plays a "leading" role in shaping mass life is in its heyday in military–corporate life, this tradition no longer commands the automatic respect it once enjoyed in left–liberal circles. The relation between planning, activism, and lived experience must be interactive as well as interdependent. People without ready access to advanced technologies and information databases create their own maps and models out of the aggressively omnipresent environment of popular or consumer technoculture, a milieu governed by the daily rhythms of work and leisure, and where alternative forecasting has little chance of playing any more than a minimal role. For the most part, that is where popular desire for a different future is learned; it cannot simply be taught from above, nor can the creation and communication of such desires be easily represented in quantitative "hard data." Alongside the need to link the data and scientific knowledge with direct action by movement activists, it is also crucial to engage with and contest the home truths with which ordinary people organize their daily lives. Much is to be gained from following the suggestion in the editorial of the first issue of *New Left Review*, which stated that the task of socialism today is "to meet people where they are, where they are touched, bitten, moved, frustrated and nauseated,"[17] and not, we might add, to tell them where they *should be*.

A similar caveat about elitism might be applied to Williams's arguments that any realistic reconstruction of an alternative future ought fully to acknowledge the limits to growth, development, and production in a world that must renew its resources or destroy itself. Today's socialist challenge, he argues, is to rethink all of these vital categories outside of the long Enlightenment tradition of scientific progress which viewed the physical world as raw material to be exploited in the interest of human society. The Marxist tradition, unparalleled in its dissent against the use of people as materials, nonetheless shared this exploitative world-view in regard to the physical world and its nonhuman inhabitants. The devastating consequences of viewing the physical world as mere raw material make it clear that no livable future is possible if current trends in capitalist production continue; they also show that a wholesale redefinition of socialist thinking is needed, one that takes into account the social costs of viewing "nature" simply as a realm of servitude from which men and women must be emancipated. What Williams calls the "constituted nature" of the physical world must be a constant and central factor in all attempts to rethink and remake the future.

The contribution of a "politics of nature" to the renewal of futurism cannot, however, be restricted to the intellectual redefinition of traditional Marxist categories any more than it can be handed over to either the traditional concerns of environmental conservators—rivers, oceans, trees, the atmosphere—or to the purificatory politics of deep ecology which venerates the necessity of uninhabited places. Any broad understanding of the "environment" and the "politics of nature" must also include the full range of issues that have come to be known as the politics of the body, the site of all modern power relations: healthcare rights; reproductive rights; sexual politics; the ethics of biotechnologies; the politics of the immune system; the politics of skin color; militarism; state surveillance; penal repression; concerns about worker safety, diet and nutrition; and so on. It is in the contexts of these environmental issues that people experience limits to their *social growth*, and these are the areas where individuals invest their strongest political passions and feel that their opinions and actions can have the most effect.

Above all, it is in the environment of the politics of the body that a progressive sense of social individualism can be posed as an alternative to the dominant laissez-faire conception of the atomized individual. Prevailing conservative definitions of the individual appeal to a limited sphere of "privacy" where our bodies are either policed by puritan morality or else sanctified by capitalism as a temple of consumer choice. By contrast, an expanded conception of social individualism involves positive, liberatory rights of freedom and choice, drawn from the guarantees of democratic citizenship rather than from the *magna carta* of consumerism.[18] Only through attention to individual rights can we build a radical democracy, guaranteeing respect both for the differences among individuals and for the environments—physical, cultural and social—that we inhabit. In this way, getting the future we deserve might still involve just deserts for all.

CHAPTER SIX

THE DROUGHT THIS TIME

The ice age is coming, the sun is zooming in
Meltdown expected, the wheat is growing thin
THE CLASH, "LONDON CALLING"

During a brief appearance at the 1990 Earth Day rally on Capitol Hill, Paul Ehrlich, biologist and ecological futurist, urged the assembled crowd and the television audience to accept that "our problems are absolutely global": "a cow breaks wind in Indonesia, and your grandchildren could die in food riots in the United States." What kind of logic was at work in this remark? Lateral thinking? Instant karma? Weird science? Under any circumstances, it sounds like a strange claim to make, almost a lampoon, hyperbolically warping whatever causal logic might link these two events. But Ehrlich judged his audience well enough; and he could count on his long experience in the ecology movement to know that most of them shared the "paradigm," scientific and political, that framed such a remark. A well-documented history of scientific research and recent scientific claims about the "greenhouse effect" filled out the picture; the politics encouraged by his speech was based on the globalist premises shared by his constituency. Given the media's widespread public airing of both the science and the politics in advance of Earth Day 1990, Ehrlich and others could hope that the remark sounded "logical enough" to pass for "common sense," or, at least, would come to resemble common sense at some point in the near future.

As an attempt to give concrete, or even proverbial, form to a developed political philosophy, Ehrlich's remark wouldn't, of course, rank as a great success. It was a little too smart, and thus too much of a shortcut for an argument that required many more causal connections to be made along the way. Perhaps its underlying meaning was still too obscure to the public mind. At any rate, the remark seemed to lack the sensuous immediacy that makes good "spontaneous philosophy" out of a mature body of political thought. This is not to say that the ecology movement hasn't produced its own effective behavioral slogans, like "Think Global, Act Local," or Arne Naess's "simple in means, rich in ends," or Barry Commoner's four laws of ecology: "Everything is connected to everything else"; "Everything has to go somewhere"; "There's no such thing as a free lunch"; and "Nature knows best." For the most part, however, they have been trite and touchy-feely—at least when compared, in terms of their rhetorical power, to appeals like "workers of the world unite, you have nothing to lose but your chains!" But the powerful appeal of ecology as a practical politics lies in its capacity to encourage people to make consistent links between the social or emotional shape of everyday actions and a quantitative world-picture of physical causes and effects. Above all, it is a politics of information and knowledge, exceptional among social and political movements in its overriding appeal to science for proof of the justice of ecological claims. In this spirit, Ehrlich's remark asks us to do a certain amount of work in making the scientific connections that support his logic. Let us spell some of these out.

It is no longer only the eco-cognoscenti who know that the flatulence from cows contains high levels of methane, one of the greenhouse gases—along with carbon dioxide (CO_2), nitrous oxide (NO_2), and the chlorofluorocarbons (CFCs)—that is believed to intensify global warming. Cattle exhaust is therefore often unfavorably compared to the capacity of trees to absorb CO_2, especially in developing countries that harbor rainforests, and where the forests are being cleared to make room for cattle ranches—in part to support the West's fast-food appetite. This cycle of commercial development is undermined by another cycle of physical effects, however. For it is widely believed that global warming, as a result of the atmospheric action of the greenhouse gases, will affect food production in certain parts of the world—especially the United States, where the climate of the mid-latitudes, the location of the nation's "breadbasket," will see prolonged droughts during the growing seasons. Consequently, food shortages and food riots are quite likely, in a country long used to the privilege of being the world's leading grain exporter, and little used to the experience of consumer scarcity for the majority, though by no means all, of its population. Following through Ehrlich's logic, then, we find that the Western narrative of exploitation and neocolonial dependency is overlaid and turned back upon itself by a narrative of retribution.

For those who do not feel implicated in the story of neocolonial underdevelopment in faraway places, the narrative of retribution will seem like an arbitrary form of justice, though it is unlikely to be reduced to the formalistic witticism so beloved of meteorologists: that the motion of a butterfly's wings in Peru can cause tornadoes in Iowa.[1] But the corollary of Ehrlich's story is one in which actions and events in North America have immeasurably more effect on people and social life in Indonesia (or Peru, for that matter) than vice versa. Automobile emissions from North America have a much greater impact on global ecology than does the cattle exhaust (itself partly determined by Western meat consumption) from countries like Indonesia. A similar irony is attached to the projected image of food riots in North America. If uneven distribution of produce and Western exploitation of Third World development have been the primary causes of scarcity in world food markets, the imagery of famine and food riots has nonetheless come, in Western eyes, to be a fixed part of the natural landscape in these distant countries. Food riots in North America are all the more terrifying a prospect because they are not felt to "belong" here; they "belong" in countries like Indonesia, where scarcity is seen as a natural result of the land's carrying capacity or as part of some Malthusian check that limits overpopulation.

However one interprets the burden of cause and influence, there is no getting away from the sure link between events in the West and events in the Third World. The logic of Ehrlich's remark is clearly intended to underscore that relationship. It is only in recent years, however, under the expanding aegis of global politics and ecological science, that this shared dependency has been brought to popular consciousness. The new consciousness, it should be noted, is not a consequence of Western guilt, but rather the result of a perceived threat to Western dominance, so dependent on resources that are now seen to be limited, or dwindling. Ehrlich's remark draws its moral authority primarily from environmental science, not from any reservoir of counterimperialist sentiment. Far from any belated recognition of the effects of neoimperialist dependency, it is none other than a set of theories from the earth science of climatology that provides the authority for the "common sense" that connects the cow to the food riot.

In this respect, nothing could be more significant than the appearance at the Earth Day rally of Bob Ryan, a well-known weather forecaster from a Washington DC television station, who called attention in his speech to the day's glorious weather, claimed the "sky" as "part of the earth," and then proceeded to scold President Bush for his inaction on environmental issues such as global warming, acid rain, and air pollution: "Mr President, look out the window!" he boomed repeatedly, in the direction of the White House. "This is what clean air looks like. This is the kind of sky we want." Here was the unusual spectacle of a meteorologist speaking as a meteorologist at a political rally, as scientists are rarely wont to do in their professional capacity. Ryan was doing more than arguing for the protection of his career commodity—fine weather. His appearance underlined the newfound celebrity status of meteorological and climatological experts within the ecology movement.

If Ryan had been up on his presidential history, he might have reminded Bush that one of his predecessors, Thomas Jefferson, had been more seriously concerned about the effects of climate change on the national economy. Jefferson, whose passionate interest in meteorology guided his personal search for agricultural improvements appropriate to the national climate, was such a conscientious weather observer that he had even stopped to pay for a new thermometer on his way to the proclamation of the Declaration of Independence! In his Meteorological Journal for the years 1810–16, Jefferson had speculated conclusively on trends in temperature statistics collected by himself and other weather enthusiasts across the country:

> It is a common opinion that the climates of the several States, of our Union, have undergone a sensible change since the dates of their first settlements; that the degrees both of cold and heat are moderated. The same opinion prevails as to Europe; and facts gleaned from history give reason to believe that, since the time of Augustus Caesar, the climate of Italy, for example, has changed regularly, at the rate of 1 degree of Fahrenheit's temperature for

> every century. May we not hope that the methods invented in later times for measuring with accuracy the degrees of heat and cold, and the observations which have been and will be made and preserved, will at length ascertain this curious fact in physical history?[2]

Despite the authority carried by the word of a meteorologist President, Jefferson's appeal to the "common opinion" that the climate had changed owed as much to popular belief and memory as it did to any evidence of statistical certainty. Popular anxiety about climate change was, and still is, an ever-present component of the US weather culture, in a country whose geographical make-up (a large land-mass that lacks any mountain ranges running horizontally across the country) produces a diversity of climates and a range of temperature extremes unmatched in any other nation-state. It is only recently, however, in the second half of this century, that climatic variability has come to be scientifically recognized as proven; only in the last fifteen years has a historical correlation been established between significant shifts in climates and concentrations of CO_2 in the atmosphere.

At the end of the 1980s, the specter of climate change, induced primarily by the burning of fossil fuels and invoked through the awesome term "global warming," had come to haunt the political soul of popular consciousness. In a short period of time, the thesis of global climate change, supported by what scientists have detected as the "signal" or "fingerprint" of the greenhouse effect, has become firmly rooted in our world-view, a permanent component of the stories we tell each other about the interplay between natural and social laws that affects our everyday life and our picture of its future. Today, in a large part of our culture, it is almost a natural part of our explanation of the everyday world. Appeals to "global warming" have become a catch-all for accounting for almost every kind of environmental anomaly. Faced with the evidence gathered in the last five years of record droughts, record rainfalls, record storms and floods, record hot and cold spells, tornado activity, and other weather extremes, it is as if the scientific ideas about climate change have taken on their material form before our very eyes.

The pathways between scientific theory and the sedimented "common sense" of popular consciousness are far from smooth. They run the gauntlets of corporate interest, political profit, and government funding. Any proper case-study of the career of theories of global warming would have to account for the thickets of influence, legitimation, and authority that acted as a filtration network for the theory. While the politics of global warming is almost wholly dependent on the word of experts, these other networks of authority and sponsorship are a powerful shaping influence on the way in which the story gets told. One result of this process, for example, is to redefine the field of science itself. Climatology, hitherto considered a second-class adjunct of the more exciting field of meteorology, or at best a branch of physics that had more in common with geography, has seen its object—knowledge about a stable archive of climate statistics—transformed into a volatile, political commodity of the first importance. Climatology is now more an analytic than a descriptive science. Climate is no longer "the average state of the atmosphere" but an unstable set of variable events, subject now to the effect of human industry. It is no longer simply affected by distant external factors like the earth's orbital tilt, the gravitational force of sunspots, or the tectonics of continental drift; today, climate is the variable outcome of internal oscillations and feedback between the major subcomponents of its system of energy exchange—atmosphere, oceans, ice sheets, land surface, and biota, of which the latter, the site of human production of greenhouse gases, has been the major focus of attention. The fragile, self-regulating economy of this climate system is now perceived to be breaking down. The cause? Excessive human intervention. The symptom? Overproduction of CO_2 waste.

It is perhaps no coincidence that this new threat is often described in terms usually reserved for the liberal market economy, and that human intervention is demonized in the same manner as "state intervention" in that economy. As one climatologist put it, "we still have to learn to live according to our climatic income."[3] Nor is it a surprise to find the moralizing burden for this interference shifted on to humanity as a whole, further Christianized by the language of retribution and penitence. As another commentator put it, global warming must be seen as "the wages of industrialization."[4] Certain elements of the new world-view that is being constructed to accommodate the global warming theory resemble pre-Enlightenment conceptions of Nature as a providential interpreter of human affairs, repaying the whole of humanity for its sins with the visiting of meteorological scourges. Other elements, of course, are more oppositional, rechanneling the blame for the gathering crisis on to the more egregious "sins" of business leaders, corporate institutions, and members of the political class.

Just as the global warming theory has served as an added ingredient for the rich stew of popular millenarianism in the last decades of the century, so too its consequences have been completely naturalized in practical folklore, in much the same way as the weather has always been used to make sense of everyday life. While the new status of the weather has made it an especially favored feature of the folklore of modern scientific philosophy, the theory of global warming itself has had to dislodge recent memories of other theories of climate change equally underscored by scientific authority. In the course of one generation—the thirty-year span that used to be the unit for measuring climatic averages—the strongest memories have been those associated with nuclear activity and global cooling, respectively. In the days of atmospheric bomb testing, cold weather was frequently blamed on nuclear fallout, not just as a result of radioactive dust, but also as the imagined consequence of radical changes wrought in the radiation belts of the upper atmosphere. Survivalist radioactive futures were a meteorological constant in science fiction. Here is a typical SF weather forecast from Philip K. Dick's *Do Androids Dream of Electric Sheep?*

> —ho ho, folks! Zip click zip! Time for a brief note on tomorrow's weather; first the Eastern seaboard of the U.S.A. Mongoose satellite reports that fallout will be especially pronounced toward noon and then will taper off. So all you dear folks who'll be venturing out ought to wait until afternoon, eh?[5]

In most projections of a nuclear winter, generated by the atmospheric effect of smoke and dust released in a major war, temperature shifts of up to 30 degrees Centigrade dwarf the single digit figures forecast by the global warming prognoses. The nuclear winter's scenario of massive crop failures, worldwide famine, and acute environmental diseases make the industrial and agricultural adjustments required to avert global warming seem like fine-tuning in comparison.

Increasingly, apocalyptic fears about widespread droughts and melting ice-caps have displaced the nuclear threat as the dominant feared meteorological disaster. In addition, nuclear winter scenarios have lost a good deal of their pictorial fascination in recent years with the waning of the Cold War. By contrast, theories about global cooling and the coming ice age have been directly contested and downrated, if not entirely vanquished, by the global warming theorists. In the space of little over a decade, the theory of global cooling, once a dominant thesis among environmental scientists, has been relegated to the margins of legitimacy, espoused today only by crackpots and conspiracy theorists who are ridiculed for their views, as if they still believed the earth is flat.

THE COOLING

As long as I can remember, we were always about to enter a new ice age. I should say that I grew up in the lowlands of Scotland, a country where history is not free of meteorologically induced hardships, but whose climate has generally been favored by the singular factor of the Gulf Stream's warming influence. Perhaps it was some subterranean effect of Calvinist determinism that encouraged me to accept that the Gulf Stream was destined to change direction, as surely it would soon, ushering in a new ice age. Perhaps another, older politics of memory about the weather was at work, recalling that a prolonged period of cold weather in the second half of the sixteenth century had helped to secure the nation's impoverishment and sealed the fate of its sovereign independence. This was a memory that qualified, if it did not directly challenge, the nationalists' more popular thesis, widely current in the sixties and seventies of my youth, that the Scottish ruling class simply "sold" the nation to England. Nationalist and theological myths aside, there was more than enough climatological support during these years for my fears about shifts in the direction of the Gulf Stream. To begin with, it was widely accepted that we were nearing the very end of an interglacial period, that climate historians had established as lasting, on average, for 11,000 years, between 90–100,000 year glacial cycles. Even within an interglacial period, it had been established that climate was by no means invariant. Comparative studies had suggested the radical effect of climate change on the fall of Mycenae, and on cultures as remote as the Hittites, forced to migrate south in the years after 1200 BC.[6] In modern times, the Northern hemisphere had undergone a series of variable climatic regimes: 900–1130, warm (heyday of the English vineyards); 1150–1380, cooler (expansion of the circumpolar "westerlies"); 1550–1850, even colder (the "little ice age"); 1850–1950, warmer.[7] During the little ice age, it is believed that the Gulf Stream had indeed shifted south, causing repeated harvest failures in Scotland. In the last few decades, since the late forties, the northern hemisphere had seen a relative cooling, and ice cover had begun to increase. The most conservative estimates of climate change during the early seventies forecast a return of the Northern hemisphere to a neo-boreal climate similar to that of the "little ice age," preparatory to the ultimate advance of the glaciers. It was suggested that a decrease of only a few degrees in average annual temperature, sustained over a number of years, would be enough to set the glaciers on the march.[8]

At the beginning of the seventies, weather culture in the Northern hemisphere, used to the relatively stable climate of the mid-century years, was transformed in response to a series of violent fluctuations in global weather patterns, which brought the disastrous drought to the Sahel countries in 1972–75, the devastating failure of the Soviet grain harvest in 1972, and, later in the decade, the first big freezes in Florida. It was a period in which country after country faced weather disasters year after year—floods, droughts, famines, food shortages, monsoon failure, and rapid increase in snow and ice cover. The effects of the OPEC oil agreements in 1974, and later the postrevolutionary curtailment of Iranian oil production, brought new meanings to the meteorologists' metaphor of climate as the world's "energy budget." With the changing face of world "food politics," a game in which grain exports were increasingly used as a political weapon, climate change was becoming a key factor; each major power initiated a national climatic forecasting program geared to short- and long-term economic planning during the mid to late seventies. In the hysterical context of this new economic situation, the weather was no longer "normal," the effects of global cooling no longer remote. Like the flood of global warming books that were to appear at the end of the eighties, a series of books by global cooling scholars was rushed into print in the mid-seventies: among the more serious, Nigel Calder's *The Weather Machine* (1975), Lowell Ponte's *The Cooling* (1976), and John Gribbin's *Forecasts, Famines and Freezes* (1977). Not long after, books with titles that hedged their bets on the future began to appear, like D.S. Halacy's *Ice or Fire? Can We Survive Climate Change?* (1978).[9] By the mid-seventies, the CIA had taken notice of the phenomenon and commissioned a series of reports about global cooling, inferring that the new "threat" of climate change was "perhaps the greatest single challenge that America will face in coming years."

To read the CIA reports is to see how efficiently *realpolitik* can convert an issue like climate change into a drama of national strategic advantage. The first report, in 1974, acknowledged that, even by the expansive standards of the CIA, the new concern with climatic issues "ranges far beyond the traditional concept of intelligence." The report established that the expansion of the US's global power had begun in the late nineteenth century, at the end of the neo-boreal period, and had thus been "blessed" with a warmer climate favorable to agricultural production in the Midwest. What was considered to be the "normal" climate of this century was in fact the most beneficial climate for this particular national economy. The second report addressed concerns raised by what was at that time an emerging scientific consensus about the reversion to a neo-boreal climate—historically, "an era of drought, famine, and political unrest." Placed on the alert by the widespread crop failures of the previous years, the United States, the report suggested, would have to reassess its role as the world's major food provider. By the seventies, largely as a result of impositions of agricultural programs by institutions like the World Bank, only seven of the world's 200 nations were net food exporters; the US accounted for three-quarters of all grain exports. The results of global cooling would be to shorten growing seasons in Canada, northern USSR, northern China, and northern Europe, and induce monsoon failures in southern China, western Africa and the Indian subcontinent. Dams and irrigation systems would be useless, the high-yield grains of the green revolution would fail; only the United States and Argentina, of the major agricultural regions, would stand to benefit from global cooling. Faced with the prospect of such an increase in power and influence, the report concludes that "the US might regain the primacy in world affairs it held in the immediate post-World War II era," but "the potential risks to the US would also rise. There would be increasingly desperate attempts on the part of powerful but hungry nations to get grain any way they could. Massive migrations, sometimes backed by force, would become a live issue and political and economic instability would be widespread." Eventually, the report falls prey to the Cold War logic of paranoia, endemic in the intelligence community. Resentment of US dominance will increase, it suggests, and "the US will become a whipping boy among those who consider themselves left out or only given short shrift. The few other nations which might have some surplus will be tempted to use it for their own political ends." The report's final scenario imagines nuclear blackmail on the part of resentful nations, allied with threats to change the climate by melting the polar ice caps.[10]

Ever loyal to its trademark paranoia, the Pentagon suspected for a while that global cooling might be a Soviet plot, consistent with the other side's plans to ruin the climate of the United States. Climate modeling at RAND and DARPA had long been applied to this scenario, while vast schemes of climate modification, some of which revolved around plans, code-named Nile Blue, to counter an alleged Soviet proposal to dam the Bering Strait, had attracted the interest and support of presidents Kennedy and Nixon.[11] At the height of the Cold War, the Pentagon was heavily involved in the use of weather modification for military purposes. One of the CIA's programs for destabilizing the Cuban economy involved an attempt to dry out the Cuban sugar crop by seeding clouds before they got to the island. Honduras sued the US for depriving it of rain as a result of a weather modification program in Florida. Eventually, the Pentagon Papers exposed the Department of Defense's costly seven-year attempt at climate modification in Laos, Cambodia, and Vietnam, including a long and focused effort to waterlog the Ho Chi Minh trail. This cast a pall over the practice of using climatic change as a weapon. An international ban was thereafter imposed by UN resolution on all attempts to influence the environment by waging climatic war, ranging from swamping by creation of tsunamis, to striking at targets with artificially induced lightning, to the destructive irradiation of selected regions by blowing holes in the ozone layer.[12] The related practice of using food as a political weapon was stepped up, however, as the US tied its food aid to the UN voting allegiances of other nations. Heinous policies of triage and "lifeboat ethics," which involved choosing among populations bidding for survival, were widely discussed, and a new ruling climatocracy came into play, exercising its sway over such organizations as the World Meteorological Organization and World Weather Watch, founded to distribute weather information more evenly and to provide needy nation-states with early weather warnings.

The strategic advantage promised to the US under the cooling regime of climatic change has been entirely reversed by the theory of projected warming conditions. In the event of global warming, it is likely that the growing seasons of northern regions like Canada, northern China, and the northern Soviet Union will be longer, while the midsections of the US will suffer widespread drought. Countries whose exporting power is not climate-sensitive, like Japan and the OPEC nations, would be unaffected. As the debate about global warming moved to the forefront of international politics in the late eighties, pressure to downplay the self-interest of sovereign nation-states became a new moral imperative of environmental global diplomacy, a move that the Bush administration had to be dragged, kicking and screaming, to recognize.[13] One can only guess at the role played in this drama by Washington fantasies of global food supremacy generated by the cooling theory of the seventies. Even more complex was the political process by which the theory of global warming came to win influence and eventually predominate over other accounts of climate change.

While CO_2 was a known factor in all accounts of atmospheric warming (as a theory, the "greenhouse effect," vital to the life-sustaining climate of the earth, was first suggested by Jean-Baptiste-Joseph Fourier over a hundred years ago), the cold earth theorists argued that it accounted for only 3 per cent of temperature variation (heating mostly in the lower atmosphere), compared to the 90 per cent caused by the cooling factors of manmade dust—from agribusiness, industry, auto and aero exhaust, slash-and-burn agriculture—and volcanic dust. More important, the cold earthers argued, global warming, a documented increase in the world's average temperature of 1 per cent Fahrenheit over the last century, was actually a factor in hastening the *end* of the interglacial period. Far from universal, warming was seen to be uneven, mostly concentrated in southern, subtropical, and low to middle latitudes. As these latitudes warm up, rising air forms moisture-laden clouds that eventually precipitate as snow in the north, adding to the ice sheets. In addition to the effects of an increased tropic–polar temperature differential, causing violent weather extremes, the increase in the albedo of cloud cover ultimately has radical cooling effects. As early as 1934, Sir George Simpson, head of Britain's Royal Meteorological Office, pointed out that glaciation required a poleward moisture transfer, generated not as a result of a decrease in solar radiation but rather by higher tropical temperatures and increased albedo. Thus, the temperature at the start of glaciation must be higher than the interglacial temperature for cooling to set in, and for polar air to move south. Under the theory of the CO_2–glaciation relation, then, a little global warming was going to lead to a lot of global cooling.

Most of the cold earth theorists still accept some version of this thesis, while the "CO_2 community," backed by government funding that has come finally to accept the warming theory as its unofficial policy, argues for the long-term prevalence of buildup of CO_2 and other trace gases as the decisive factor in the debate, over and above the "local" significance of factors such as cloud albedo increase. Consequently, the global warming thesis became the dominant scientific theory in the early eighties, and has taken on a factual status in public consciousness in recent years. A decade of very strange weather helped to solidify popular acceptance of the warmers' claims. Weather anomalies in the eighties included: the six warmest years of the century (1990 being the warmest in recorded history); the El Niño/ Southern Oscillation of 1982–83 that rocked weather patterns everywhere and left hundreds of thousands dead in the wake of massive floods, coastal landslides, dust storms, cyclone blasts, record rains, and mass migrations; and, of course, the annus mirabilis of 1988, the year, when, in President Bush's words, "the earth spoke back," giving millions an idea of what global warming must surely feel and look like, with a devastating drought in the Midwest and Plains states, enormous forest fires all over the West, the mighty Mississippi almost dried up, withered crops, slaughtered cattle, virtually uninhabitable urban centers, and, to top things off, the ravages left by Hurricanes Gilbert and Helene in the Caribbean and the Great Flood of Bangladesh. The anomalous weather of 1988 was due, for the most part, to a split in the jet stream in the spring that prevailed throughout the summer. But with the extremity of that summer felt by everyone, there were few who were not tempted to reach for the global warming theory as a ready explanation, while scientists who stuck their necks out to provide authoritative support for the connection between warming theory and practice were rewarded with the heady oxygen of media publicity. Others remembered that a series of viciously cold winters in the mid to late seventies had provided similar, experiential support for theories of the "coming ice age."

In the face of the new scientific consensus, cold earthers stood their ground. Some, marginalized as conspiracy theorists, continue to cite the warming theory as a government cover-up influenced by the short-term interests of the energy industry (funding for the CO_2 debate comes from the Department of Energy not the EPA). They pointed out that the projected long-term effects of warming involve much less in the way of immediate business regulation than the more urgent action called for by the cold earthers, whose disaster scenarios of glaciation stretching down to New York City are projected for the very near future.[14] Of course, it has not helped their case that projected dates for the great cooling have come and gone—most notably in the case of John Hamaker, the autodidact and experimental farmer much lionized in cold earth circles, who forecast that the end of the interglacial would commence in 1990.[15]

BALANCING THE BUDGET

When it comes to actions and reforms relating to environmental protection, the urgency of the necessary remedies may vary, but there is little to choose between cooling and warming. Much the same causal factors are at stake, and both require similar preventative measures on the part of industry and personal consumption: reduction of CO_2 and other trace gas emissions, including CFCs (which are also ozone-depleting); radical regulation of industrial pollution; protection of forests and massive revegetation; restructuring of energy uses aimed at an immediate shift in the energy base; and a transformation of the high-consumption lifestyles of most Western citizens. Mindful of its mission to safeguard the profits of the energy and utilities industries, the Reagan administration recognized the common consequences of both theories by slashing government funding for any kind of CO_2 research—although its environmental policies will best be remembered by Secretary of the Interior Donald Hodel's comment that people should respond to ozone depletion by wearing baseball caps and sunglasses. In recent years, it has become more difficult for central government to justify its inaction by setting one theory off against the other; in other words, by opposing the coolers' case against manmade dust and particle pollution to the warmers' case against greenhouse gases, as if the respective cooling and warming effects will simply cancel each other out. Nonetheless, the continued existence of debates on climate change (especially around the variability of feedback factors such as clouds, oceans, and biomass) bolsters an air of uncertainty that justifies the Bush administration's moratorium on action, just as its predecessor had deferred making national policy about acid rain. Presidents fiddle while the globe burns.

In the meantime, ecologists and biologists have reminded us that what lies in the balance is the extinction of millions of species as a consequence of extreme variability in weather patterns, the unrealized effects of the greenhouse gases still en route to the stratosphere, or the prospect of a chain reaction whereby melting permafrost in the tundra releases vast quantities of methane from ground-locked biomass compounds. The speculative calculus of disaster, endlessly computable into new configurations, is a favorite game of environmental scientists and corporate-minded ecologists involved at the advisory and lobbying levels of the debate. Manipulations of this calculcus have brought the physical world firmly within the purview of technocratic futurology. The experts' models of global warming present a complex, interactive picture of feedback components, with projected statistical effects from one sector—cloud albedo, or ocean absorption of solar radiation—played off against another—CO_2 production from rotting vegetation, or the role played by marine phytoplankton. This modeling is governed by the new corporate logic of planetary management, with its centralized rationalization of the climate system's every conceivable component. The same cost–benefit logic is evident in new forms of global economic management, with its debt-for-nature swaps, and the growth of an international market in tradable pollutant emission rights. While these developments are clear evidence of the political and economic impact of ecology at the highest levels of decision-making, there are reasons to be wary of a distributive system with such an Olympian perspective.

The consequences, on the ground, of this eco-mercantilism are often quite dismal. The more global the model, the more likely that attention to the social causes and conditions of the climate crisis will drop from view. We are left with a formal calculus of the world's "energy budget" (it must be balanced) or "climatic income" (we must live within our means). Whether this involves calculating the net insurance premium to be paid by "us," or estimating the gross global deficit—excessive "total respiration" of the biosphere measured against total photosynthesis—the implied aim of restabilizing the planetary economy is to ensure that the "economic climate" of the earth's resources is a favorable one for the future of business. While meteorologists have long sought to advance their profession by selling weather-related services to industry, only in recent years, with the advent of fears about global climate change and ecological degradation, has the language of climatology become a privileged vocabulary for futurological business forecasting.[16] In this respect, the emergent rhetoric of economic climatology reflects anxieties about global crises for climate and for capital alike. Unlike the "weather," which is still locally variable and thus a risky investment, the "climate," once assumed to be regionally stable, is now seen as a global system in a state of uncertainty. The implied solution, then, is to "compete" with nature in (re)creating a favorable climate by balancing resources against expenditure.

Although my comments thus far refer primarily to trends in language use, there is no easy separation, here, between metaphor and action. This budgetary way of looking at the world—fueled, as I have suggested, by anxieties generated by the global warming debate—is continuous with the scientific perspective of quantitatively dominating the physical world. Now that science has shown the clear impact of the "human fingerprint" on a global system so vast as atmospheric behavior, such a logic demands the more stable, guiding influence of a whole hand. If humans are now *competing* inevitably with nature in the fight for a stable climate, they need to win. Such a logic, in other words, demands that attempts be made to control the interaction of the various components of the climate system: oceans, ice sheets, land surface, atmosphere, and biota. As the economic scope of capitalism enters into its truly global phase, it is clear that this logic of reckoning inputs against outputs is entirely complicit with the interests of the new global investors, a spectrum that runs from small-time players on the futures market, so heavily determined by the effect of local meteorological fluctuations on food commodities, to the earth-movers and shakers at the World Bank, whose efforts profitably to shape the future of the multinational economy are equally dependent on regional climatic stability. According to this logic, all attempts to "deregulate" the climatic economy (in this case, reducing the influence of human industry by curtailing greenhouse emissions and replacing the source products with substitutes) *must* also be seen as opportunities for regulating the physical world that did not exist hitherto. Greater powers of regulatory control are thus claimed in the name of allowing the system to revert to its "natural" self-regulating economy. This is the contradictory form in which laissez-faire economics have been advanced throughout modern capitalist history. If we fail to see how this logic of regulation/deregulation carries over into the claims currently being made in the name of corporate ecology, then we fail to grasp the full significance of the debate about "global warming" that has come to occupy center stage in world politics so soon after the breakup of the fixed Cold War order. The crusade to claim the whole world as "free" for liberal capitalism is currently locked in step with the campaign to "free" the climate from human influence. History suggests to us that both definitions of "freedom" are shot through with the lowest form of irony.

Some of this irony's antisocial force can be brought out by considering the obverse of the logic of human control over planetary management, in the theory that has come to be known as the Gaia hypothesis, first proposed by the English scientist, James Lovelock. According to the Gaia theory, organic life is an active (negative) feedback component of the planetary system; therefore interaction between organic species is one of the features that helps to regulate the system's homeostatic economy. Gaia's adaptive control system includes its self-regulation of climatic and atmospheric composition at optimal levels. What matters most in Gaia's living planetary system is the maintenance of Gaian life, not human life. When the earth's self-regulating organism shifts into a new stable state—global warming, in this case—in order to protect itself, it is likely that conditions are created that will not be favorable to the continuance of human life. Global warming, which might entail the elimination of humanity, an unhealthy species, is thus the planet's "solution" to its ecological crisis. Since human life has proven to be the chief threat to the health of the planetary organism, it is in the Gaian interest to eliminate human life. Gaia, sadly personified by Lovelock as the earth goddess:

> is no doting mother tolerant of misdemeanors, nor is she some fragile and delicate damsel in danger from brutal mankind. She is stern and tough, always keeping the world warm and comfortable for those who obey the rules, but ruthless in the destruction of those who transgress. Her unconscious goal is a planet fit for life. If humans stand in the way of this, we shall be eliminated with as little pity as would be shown by the micro-brain of an intercontinental ballistic nuclear missile in full flight to its target.[17]

While some deep ecology supporters of the Gaia thesis see it as an effective philosophical myth for countering the chauvinistic logic of human self-interest or as a vehicle for proclaiming the "liberation" of nature, critics see it as a form of macho environmental fascism that necessarily favors the good of the state/earth over the good of social groups and individuals. The Gaian thesis simply inverts the logic of human domination over the natural world: planetary management is seen not as an extension of human control, but as a process to which the fate of humans is utterly subjugated. Under cover of the rhetoric of "biocentric equality" and the "balance of nature," the logic of domination is held intact, and the social specificity of human life drops out of the picture.

Like global models of corporate planetary management, which take the planet as an economic unit, Gaian philosophy demonstrates the danger of taking the planet as a zoological unit. In either case, humanity appears as a mythical species, stripped of all the rich specificity that differentiates human societies and communities, and oblivious to all the differences in race, gender, class, and nationality that serve to justify and police structures of human domination within and between these societies.[18] In both instances, the questions raised by ecology can no longer be explained or answered by social theory or social action; they are resolved at the level of "resource management" by the logic of the multinational corporate state, or by the independent diktat of the "tough" planetary organism. The problem of global warming is no longer an arena for exposing the barbarism of social institutions. From both of these perspectives, it is an experimental opportunity to test the logic of their respective world-views. It is not surprising that a recent book, *Gaia: An Atlas of Planet Management*, has succeeded in combining with ease both philosophies, celebrating the adventure of safeguarding Gaia's health through macromanagement of her resources.[19]

To see how this experimental adventurism is shared by the scientific establishment, we need look no further than the words of Roger Revelle and Hans Seuss of the Scripps Institute of Oceanography. In a 1957 paper about CO_2 exchange between the ocean and the atmosphere, often cited as the origin of the modern debate about global warming, they write:

> Human beings are now carrying out a large scale geophysical experiment of a kind that could not have happened in the past nor be reproduced in the future. Within a few centuries we are returning to the atmosphere and oceans the concentrated organic carbon stored in sedimentary rocks over hundreds of millions of years. This experiment, if adequately documented, may yield a far-reaching insight into processes determining weather and climate.[20]

It is possible that a reader could fail to detect the note of warning in these words. But there is no mistaking the euphoria attached by the authors to the very idea of such a grand experiment. The first and last experiment in which modern science can take the entire planet as its test object! Surely this qualifies as a culminating moment in the history of natural science, whose founding proponents chose the physical world as their experimental object to master through the force of rational explanation. While it is important to take note of such an experiment's sublime lure (reprised in a 1986 NASA report—"we are conducting one giant experiment on a global scale"), we should also take care not to aestheticize its attractions further. To see the ecology of global warming as an opportunity for an "experiment" is a profoundly political way of seeing the world that undercuts our best hopes for reclaiming the environmental sciences as an ecological ally.

When I describe this view as political, I do not mean that such an "experiment" would always be framed and influenced by explicit ideological aims, although there is no question that external politics has played a significant role for scientists in the climatic debates over the year. Government funding, career prestige associated with the winning theory, the opportunity to testify, lobby, formulate and adminster policy, and the advancement of the discipline of climatology are only a few of these political factors. Nor do I simply mean that such an "experimental attitude" naively assumes the possibility of objectivity for its observer. The debacle of the "hole" in the ozone layer, undiscovered for so many years because its observers programmed their computer to ignore measurements that diverged too greatly from expected norms, notoriously proved how highly "interpretive" such climatic experiments can be. Rather, what I mean is that the experimental attitude, especially when it takes the whole planet for its laboratory, becomes a form of constructive power that reshapes the world in a different image, detaching it from meaning and value and delivering it up to the rationality of technical description and control. It has often been argued that the goal of natural science, in its intention to construct a world of fact, has proceeded apart from nature, and only in relation to its own intellectual paradigms and self-defined modes of rational inquiry and verification. Historically opposed, by its own founding principles, to the local influence of political coercion and supernatural faith alike, the natural sciences have developed in relation to a much more powerful ideology bound up with the rational organization and domination of nature. Viewing segments of the natural world as controlled experiments is one of the normative instruments of this kind of rationality. The modern history of environmental policy bears the impress of this way of thinking, whether in the traditional conservationist mode, with its protected enclaves of "nature"—wildlife refuges, zoological museums, national parks, and wilderness areas; or in the more technocratic mode of "resource management." With the maturing of the idea of global ecology, publicly dramatized by the global warming debates, the entire planet becomes a protected preserve, the object of an experiment in which the global ecosystem must now be managed and regulated to sustain its organic life-forms and other "resources." The difference is that there is no way of claiming an "outside" in such an experiment, no value-neutral perspective for its observers and supervisors, and no surrogate point of view available for nonexperts who are not part of the intellectual conversation about its outcome.

If this new stage of monitoring a planetary ecology does represent a qualitative shift in the "project" of dominating nature, then it goes hand in hand with the new forms of subjectivity being forged by an emergent global politics. Concerns about planetary survival are a crucial part of what it means to be a global citizen, to have a global world-view, and how that is going to affect everyday life in places as far apart on the development spectrum as Los Angeles and Upper Volta. We live in the early formative days of globalist ideology, and, as a result, are perhaps well placed to see the full extent of how unevenly developed this ideology and its effects are. Today's debate about global warming, in the forefront of the highest diplomatic discussions among Western industrial powers, can be read as a symptom of the way in which many similar decisions will be made about global matters in the future. The fact that warming across the globe is highly uneven might alert us to some of the geopolitical factors involved. It has been suggested, for example, that the "cooling" theories of the seventies were based on "local" North American weather statistics, and therefore did not accurately reflect global trends. Evidence now suggests that the warming, if it is "global," is not at all uniform, and that the southern hemisphere and the tropics seem to be warming faster than northern regions.[21] Whether these claims prove true or false in the long term, they are not the kind of claims that play well to Third World nations, justifiably suspicious of Western theoretical science that declares universality for itself. Much of climatic science is based upon records from the northern hemisphere, the exclusive source for almost all of its historical data. The northern view, along with theories that most affect the northern hemisphere, tend to be normative in the history of modern science. As it becomes increasingly capital-intensive, science is exclusively based in the West. Add to this the warranted suspicions of developing countries like India and China (which tend to outweigh the Alliance of Small Island States, whose lands may disappear altogether under a rising sea) that the anti-greenhouse measures which have met with international consensus will have the effect of limiting the growth of their economies in order to protect the shorelines of the most developed nations. The result is part of a familiar logic. No one needs to doubt the urgency of the greenhouse problem to recognize that any Western suggestion of standards for the development of other countries is also a reinforcement of the long history of colonial underdevelopment of the non-European world. It is in such a context, and with such a historical backdrop, that one could justly say, "you don't need a weatherman to know which way the wind blows."

SCIENCE AS CULTURE

Western climatology may be recognized as the primary authority in defining the exact shape of the global crisis, but there are a host of questions related to the climate debates that touch upon local and cultural, rather than global and scientific, interpretations of the weather. Here, we come across the vast spectrum of cultural differences in living with and interpreting the physical world that have little to do with the "universal" claims of global climate modeling. Even in a local context, some of these differences are overtly cross-cultural. For example, Mary Douglas has described two neighboring tribes in the Congo—the Lele and the Bushong who live on different banks of the Kasai river—who experience the same climate quite differently, celebrating their hot and cold seasons at opposite points of the calendar. Meteorological records kept by Belgian authorities showed that there was little "objective difference" to account for each tribe seeing the other's cool season as unbearably hot, and vice versa. Douglas concluded that the phenomenological differences could be explained by the respective agricultural timetables of the tribes.[22]

Then there are the complexities of linguistic translation across climates; it is pointless for example to make much sense, in any subtropical culture, of the phrase "Now is the winter of our discontent." In addition, these problems in translation are habitually burdened by ethnocentric assumptions. For example, Jody Berland has pointed out that we tend to see the Inuit's famous multinaming of snow in "terms of an objective, functionalist nominalism more or less parallel to our own, not in terms of any fundamentally different spirit of naming." Our system of naming snow relates to the degree to which it will restrict our everyday business. In contrast to our regulatory naming of snow by "measuring its nuisance value," Berland points out that any other way of "naming" snow is seen as "a childish inability or refusal to subjugate weather."[23]

But it is not just Third and Fourth World cultures, traditionally seen as "closer" to nature, that introduce interpretive "noise" into the experts' weather control systems and climate models. Each northern nation and, more often than not, each region has its own weather pathology, richly seeded with popular memory and historical lore. One region's "warm" is another's "hot": one season's "cool" is another's "warm." For anyone, for example, who grew up in an agricultural region given to periodic drought, the prospect of long, soaking rains could never have the negative value associated with its appearance as a blight on some urban region's weather.

To acknowledge this welter of interpretive variations, crisscrossed by regional and cultural differences, is to acknowledge, of course, that we don't all share the world in the same way. Despite their inbuilt appeals to universal truth, most weather proverbs are correct only 50 per cent of the time (many carry the facesaving qualifier "oft," but, in most cases, you can usually interpret them to mean the opposite), and are otherwise appropriate only to certain local conditions and climates. People are likely to have allegiances to local cultural formations and identities before they recognize global appeals to their attention. In many regions of North America, a tradition of local pride in "state weather" is imbued with the fierce political legacy of "states' rights." Consequently, there is a long and well documented history of deep-rooted suspicion of centralized weather forecasting, especially when it involves experts reading a computer screen rather than scanning the skies or reading the signs in nature. For many years after the inception of daily national forecasts, local newspapers preferred to publish the forecasts of the local weather prophets, often very colorful characters whose community standing had been challenged by the new professionalism. The history of forecasting is marked by notoriously incorrect prognoses on the part of official meteorologists that were seized upon to fuel local skepticism. Provincial forecasters have, in recent years, acknowledged this resentment and skepticism by drawing upon the observations of local observers and spotters, who report weather variations from their home stations all over the outlying region. The weather, after all, is described and defined by millions of opinions, but the professional's forecast overpowers all other definitions. Not surprisingly, people often feel that professional forecasters, especially those in a centralized national bureaucracy that relies heavily upon computer modeling, are too remote, geographically and culturally, to do a proper job.

A good deal of cultural power rests upon the maintenance of that centralized authority in times of natural emergency and disaster, and even more rests upon the exercise of authority across cultures with unequal access to scientific information. One of my favorite examples is the description in Zora Neale Hurston's *Their Eyes Were Watching God* of

Florida's great Okeechobee flood of 1928, caused by a hurricane that the official meteorologists did not predict until the night before but the Seminole Indians foresaw in good time, thanks to their interpretation of an unseasonable blooming of saw grass. Whole sectors of the animal population joined the Indians in a forced march away from the Everglades, watched over uneasily, in Hurston's fictional version, by the mostly black workers on the swamplands:

> Some rabbits scurried through the quarters going east. Some possums slunk by and their route was definite. One or two at a time, then more. By the time the people left the fields, the procession was constant. Snakes, rattlesnakes began to cross the quarters. The men killed a few, but they could not be missed from the crawling horde. People stayed indoors until daylight. Several times during the night Janie heard the snort of big animals like deer. Once the muted voice of a panther. Going east and east. That night the palm and banana trees began that long distance talk with rain. Several people took fright and went in to Palm Beach anyway. A thousand buzzards held a flying meet and then went above the clouds and stayed.

A fellow worker, informed by his uncle of the official hurricane warning that is finally posted in Palm Beach, tries to persuade Tea Cake to leave:

> "De Indians gahn east, man. It's dangerous."
>
> "Dey don't always know. Indians don't know much uh nothin', tuh tell de truth. Else dey'd own dis country still. De white folks ain't gone nowhere. Dey oughta know if it's dangerous. You better stay heah, man. Big jumpin' dance heah, when it fair off."[24]

As it happens, Tea Cake may have been right about the Seminoles. On certain famous occasions, such as this one (and again, in 1944), they proved superior to "white science" in their prediction of the behavior of Florida hurricanes; on many other occasions, however, their homeopathic methods of reading nature's signs have been less successful than the Weather Bureau. But, in view of the disaster that the Okeechobee flood visited upon the workers and local population (up to 2,000 dead), Tea Cake's explanation of the knowledge/power relationship between the native and the white Americans has an especial irony. In particular, it positionally describes his own people—African–American agricultural workers—as a culture no longer in touch with the "signs" of nature, no longer living "in" the earth like the Seminoles, but subordinated now to the authority of another culture, whose scientific understanding of nature has underlined its ascendancy in the world. Hurston plays this unhappy positionality to its full tragi-comic effect.

A striking contrast to Hurston's coy ethnography is Saul Bellow's *Henderson the Rain King*, a rather condescending narrative about a wealthy white American's absurdist adventures in "darkest Africa." Entering into a wager with the king of the drought-plagued Wariri tribe over the likely success of their rainmaking ceremonies, Henderson impulsively takes part in the ceremony and is ritually accepted as the tribe's "rain king" when his actions are succeeded by rainfall. To play his role as rain king, it doesn't seem to matter that he is a foreigner who has had no functional standing in the tribal culture hitherto. Unlike the nearest equivalent office of "rain king" in his own culture, which would require the authority that comes with accredited learning and professional meteorological expertise, Henderson simply accedes to this important tribal position because he is in the right place at the right time. His accumulated knowledge about how rain is formed, limited as it may be to theories culled from his wife's subscription to *Scientific American*, has absolutely no bearing on his important function in the business of rain formation among the Wariri.

These stories about the relation between cultural power and climatic prediction do not *seem* to be part of the same interpretive system as, say, the eminently scientific theory of global warming, and yet it could be argued that the only difference is that they appeal to differently organized systems of rationality. Global warming theory claims universal scientific truth for itself, against which climatic interpretations like those of the Seminoles or the Wariri are seen as local belief-systems, or, at best, *ethnometeorology*. The distinction, however, is itself an exercise of cultural power. Global warming theory is nothing if not a high cultural expression of Western science, dominant in the field of interpretations of the climatic economy. But one does not have to speculate about what the Wariri would have made of the theory of global warming to recognize the local nature of a world-view that Western science poses as universal. Nor do people need the universal comfort of science to draw their own culturally coded conclusions about human pollution of the environment. As Claude Lévi-Strauss has shown in *The Origin of Table Manners*, many of the tools and rituals of everyday life—wearing hats and gloves, drinking through straws, using combs and utensils—are ways of regulating our exchanges with the external world, protecting the environment from our own polluting actions by ensuring "that nothing should be brought about too precipitately."[25]

Relatively untouched by the precision language of the meteorologist, most people in the industrialized West have their own ways of making sense of the determining role of climate in their everyday lives: from the empirical end of the spectrum—the vestigial memory of weather folklore, and the habitual practice of phenology, "the science of appearances," by which people drew seasonal conclusions from the signs of nature (appearance of flowers, thaws, animal and bird migrations, and the like)—to the more abstract domains of regional weather pride, the rule of seasonal expectation over cycles of endurance and pleasure, disaster culture, etc. This broad spectrum, and the conception of the world that it supports, has little to do with the interactive global model of atmosphere—oceans/ice sheets/cloud albedo/biota—that constitute the climatologist's weather system. In some cases, the language is different, but the lived object of experience/knowledge may be the same. For example, the infamous Southern Pacific ocean phenomena that are known colloquially as the Christ Child—El Niño and La Niña (warm and cold respectively)—take their name from the fact that their climatic influence over Central and Latin America begins, in active years, towards the end of December. The meteorological term for this seesaw pattern of rising and falling barometric pressures in two large regions of the Pacific is the Southern Oscillation. The popular connotations of El Niño in Latin religious history give the extreme weather associated with the phenomenon a cultural resonance quite different from that conjured up by the much less colorful term, Southern Oscillation.

But the difference between the language of scientific expertise and the popular culture of experience and local memory is not a difference in kind. Nor, if what is at stake here are different conceptions of the hold the physical world has upon us, can these differences be explained away as cultural relativism. They are precisely ranked on the scale of power, and they reflect real inequalities not only in the degree of power that different cultural groups have over their relation to the physical world, but also in each group's ability to make arguments that will affect that relation. In this respect, it is a mistake to think that theories of climate change that take the globe as an experimental object can rest simply upon criteria of empirical verifiability or formal coherence. These theories draw their power in the world from an elite culture peopled by those accustomed, by education and an inherited sense of entitlement, to see the globe as part of their *dominion*, a territory that exists to be rationally surveyed, itemized in a cost–benefit analysis, and protected by political action that further regulates its natural economy.

In making these points, it should be clear that I am not, of course, *disputing* the theory of global warming. To dispute a scientific theory involves vast amounts of laboratory capital and a long history of professional prestige in a specialist discipline. Rather, I am calling attention to the cultural and political conditions under which contests for theoretical dominance take place, and scientific "common sense" is subsequently shaped in the public mind. Nor am I arguing that the environmental consequences of "global warming" should not be acted upon in quite radical ways. Whether the hypothesis of global warming is proven or not, the recent spotlight on the climate debates has provided the single best opportunity for ecological condemnations of capitalist growth and development to win a hearing in the most powerful circles of decision-making. Consensus about this "crisis" will lead to significant ecologically minded steps and programs at all levels of action, from individual consumer choice to multinational regulation. So too, the need to discourage thinking about nation-states as climatic islands is concomitant with the need to construct a global politics that transcends the isolated obsessions of individual nation-states.

But we cannot expect all the changes to be progressive. Globalism will generate new power relations as the old national allegiances lose their sway. Just as the social costs to capitalism of environmental regulation are likely to be internally absorbed and handed on to consumers, so too there are cultural costs to be borne in transforming people into global citizens. This goes well beyond the tendency, already well advanced, of passing on to individuals the deeply moralistic sense of assuming responsibility themselves for the very largest ecological problems that ought to be borne primarily by corporate executives and their stockholders. For most individuals, the new "natural" scarcity of a globally conscious economy will be exploited in much the same way as the old "artificial" scarcity was exploited in the "national interest" of a few. This means more than a continuation of deprivation for many in the name of those few who benefit from creating conditions of scarcity. In cultural terms, global consciousness also means an erosion of diversity, a flattening and incorporation of cultural differences that the new "global economy/climate" can no longer "afford" to sustain—except, of course, in the postmodern realm of images, as representations of cultural difference à la Benetton. What is being erased are all those features of cultural difference that cannot be readily translated across cultures into the visual language of the global village, the polyrhythms of world beat, and the ideology of market pluralism. As the race for global culture quickens, we must be prepared to make the same arguments about cultural diversity as ecologists have made about biological diversity. The attrition of cultural diversity, like the loss of life-species, decreases the chances of sustaining our social survival.

As a world-view, globalism will have to work hard to produce a new sense of cultural allegiance, a new sense of subjective loyalty. Where the divisive course followed by the history of nation-states provides little support for expressions of global subjectivity, the new planetary dimensions of environmental responsibility are already providing the language of necessity required to pull together what has been, until recently, the disparate Babelian vocabularies of global citizenship. The move in this direction is double-edged if one considers the new range of responsibilities to be explored and exploited. Conflicts over the rights of global citizenship are unlikely to center upon the same rights of freedom and justice that have marked the bloody history of citizenship in liberal nation-states. Everyone, after all, *belongs* to a global society. No one, in principle, can be excluded; least of all "indigenous" peoples historically denied full citizen rights in liberal nation-states (although the outlaw class of "international criminals" or "terrorists," even now being extended to whole peoples like "the Arabs," would be first in line to be denied such global rights, a tendency foreshadowed in the Gulf War, the first holding action fought in the name of the "New World Order"). But the extension of citizenship in the name of laissez faire-ism will also provide further opportunities to extend liberties only at the expense of equality. Bear in mind that the globalist move is not taking place in a political vacuum. Far from being forged in the spirit of internationalism espoused by the more enlightened socialists, globalism is advancing under the aegis of free market ideology, just as environmentalism is increasingly posed as a social cost to be borne equally, by all individuals, rather than by its primary corporate and institutional offenders.

The debate about global warming, as I have discussed it here, promises to generate similar contradictions in the way people think about the natural world. Instead of feeling the weather as we have felt it historically, as part of a shared local, or even national, culture, we are encouraged to think of it globally. On the one hand, this will promote our attention to the interdependency of environmental factors across cultures and continents; on the other, it may help to dissipate people's faith in the efficacy of local actions. It has become convenient, for example, to cite global warming as a distant, almost inevitable, causal explanation for a range of environmental problems and issues with a much more local provenance, accountable to pressure by local communities and open to change by local action.

Global weather culture may be in its infancy, but the phenomenon of weather cultures was not born yesterday. In the pages that follow, I will describe some of the historical features of local, more specifically national weather culture that have served to maintain pre-global ideology until now. In this history, we will find clues and tendencies, rather than iron laws, to suggest the future's possible shapes.

PLUVICULTURE

A student of mine once revealed that he never consulted weather forecasts. "The weather is for other people," he announced, by way of explanation. I responded by suggesting that I didn't think this was a very citizenlike thing to say. Having thought about that reply for some years now, I recognize now that I was both right and wrong in responding the way I did. The weather is not simply for other people, because we share other people's weather. On the other hand, the history of national weather, as I shall outline it here, holds no simple or single guarantees about what it means to be a responsible weather citizen.

In 1841, at the Académie Française, Louis Arago, a scientist and a gentleman, who no doubt loved to make such pronouncements, observed: "France has its Cuvier, England its Newton, and America its Espy." For those who have never heard James Pollard Espy, the most colorful of fledgling meteorologists, there is much to be learned from his life and work, especially in the days of global warming, about the relation between science, ideology, government funding, and the popular imagination. Author of a polemical volume called *The Philosophy of Storms* (1841), he became a regular lecturer on the lyceum circuit, where he was known and heralded in newspapers as "The Storm-King." To be a success in this capacity was no mean achievement. Of the popular lyceum stage, Thoreau said: "Men are never tired of hearing how far the wind carried men and women, but are bored if you give them a scientific account of it." And yet this is precisely what the celebrated Espy did. For over twenty years, moreover, he was involved in a very public debate with the eminent scientist William Redfield about the nature of storms. Both proclaimed universality for their own laws of storms; both produced rules for mariners to negotiate storms; both entered into conjecture about the nature of the weather in the "far west," beyond the Alleghenies, and thus the borders of the known world; both converged on site in the aftermath of great storm damage in order to study its effects; and both drew entirely opposed conclusions about all of these subjects.

Espy's scientific theories have had mixed reviews. His deductions about storm behavior proved to be wrong; he argued for the elliptical nature of winds around a low pressure system, while Redfield argued for circular wind behavior. On the other hand, Espy is commonly credited with important discoveries about thermal air currents, especially the role of hot rising air in the generation of thunderstorms. Indeed, his notoriety rests upon one particular version of this thesis. For many years, he advertised to the public his theory of artificial precipitation, or rainmaking, based upon his observations that the thermals rising above urban manufacturing areas often produced rainclouds. In *The Philosophy of Storms*, he cites as evidence of his theory the testimony of one Benjamin Matthias from Philadelphia:

"In the course of the last winter, while in England I visited Manchester four or five times, and on each day it rained. Several of the inhabitants assured me that it rains in Manchester more or less every day in the year."[26] While Espy acknowledged that this theory sounded like "bad philosophy," he warned that popular opinion will nonetheless show it to be truthful, no doubt by accumulating testimony of the sort provided by Matthias.

As chairman of a committee reporting to the US Congress on meteorological matters, Espy was in a position to push his views further, especially at a time when rainmakers' services were often the basis of a livelihood.[27] In fact, as a measure intended to alleviate the prevailing drought-like conditions in Pennsylvania and the Mid-Atlantic States, Espy proposed that areas of forest land be burnt in order to precipitate rainfall (a proposal that could only have been considered at a time when vast areas of forest were being cleared from the land). The success of his experiment, moreover, was to be consequent upon a considerable financial reward from the government: on a scale, from $5,000 for 10 square miles of precipitation to $50,000 for the feat of keeping the Ohio River navigable in the summer for steamers.

Aside from a good deal of ridicule, in and out of Congress, nothing much came of his proposal, although at least one populist Senator saw quite clearly that no single individual, no matter how gifted or visionary, ought to have such sovereign control over the Ohio River. John Quincy Adams, the ex-president, had peculiarly harsh words to say about the ambitions of the "storm breeder": "The man is methodologically monomaniac and the dimensions of his organ of self-esteem have been swollen to the size of a goiter by a report of a committee of the national Institute of France, endorsing all of his crackbrained discoveries in meteorology."[28] Despite the ridicule and the calumny, Espy turned out to be quite correct about the capacity of rising hot air to affect the weather locally. Indeed, Matthias's observations about rainfall over Manchester have long been borne out: there is a noticeable decrease in precipitation over large industrial areas at weekends, when factory smoke and auto exhaust are diminished. Espy's forest-burning proposals for rainmaking, however, will seem like bad historical faith today in the light of the disastrous consequences of the deforestation of the Amazon, the Sahel, and the slopes of the Himalayas. Two thousand years of clearing forests for profit, reinforced by the

Christian logic of destroying groves sacred to pagan religions, have had a devastating effect on regional climates, and may now be playing a major role in global warming.

The "profit" that Espy proposed to Congress for his rainmaking scheme in the Ohio Valley was tied into a specifically national system of value. His final court of appeal was based on the assessment that the national benefits to industry and agriculture resulting from this artificially produced rainfall would be generated for the sum to each citizen of less than half a cent a year.[29] This proposal might be read as an early example of the theory of the popular distribution of social costs, borne for resources like water that can no longer be provided free to industry and agriculture. In addition, Espy's appeal links a national economy of value with what could be called a natural economy of value. In doing so, we could say that his appeal exposes the underpinnings and workings of ideology itself, inasmuch as ideology can be defined as that kind of discourse which conflates these two economies, presenting the social as if it were natural. Nationalist ideology depends explicitly on the conflation of these economies. In looking at the history of "national weather" in the pages that follow, I will consider some of the languages, technologies, and systems of representation that have helped to establish this intimate relationship as if it were a part of manifest destiny. In doing so, we will no longer be talking about the history of "weather" in the physical world; we will have to talk about the history of "the weather" in the ideological world of a national culture.

Espy's place in that history is indisputable, if only because of his rainmaking theories. His was an age of nationalist expansion westwards, governed by Jefferson's agrarian vision, and, at a time when there was a good deal of debate about the effect of land clearing on climate, his rainmaking theories drew upon the popular perception that "rain follows the plow." In other words, as the frontier pushed westward, it "created" the weather at one and the same time, modifying the climate and producing precipitation for the plains, newly under cultivation. The "national weather" was, in effect, exactly coterminous with the ever-expanding boundary of the nation's body. In this respect, we can speak of a "national" weather culture, however imaginary, that corresponds to this first phase of national expansion.

But Espy's place is also assured on account of his attempts to amass and collate on a national scale the extensively kept daily weather records of private citizens—a great American tradition established and religiously observed by Washington, Jefferson, Franklin and thousands of other unsung weather diarists. A good contemporary example of this tradition was Charles Peirce, who, in 1847, was urged "at the friendly solicitation," as he put it, "of a very considerable number of highly respectable citizens of Philadelphia," to publish a compendious record of weather events and temperatures for the fifty-seven years between 1790 and 1847. To this meteorological record, Peirce appended a number of often extensive historical lists: cold and stormy winters in Europe from the dawn of the Christian era, and in America prior to 1790; storms and hurricanes all over both continents; notes from his records concerning "the formation of the government of the United States"; a detailed history of Philadelphia, its buildings and communications—railroads and steamboats, and the concomitant damage inflicted upon them by the weather; a record of large fires and remarkable earthquakes; and an account of the history of North America from the voyages of Columbus onwards, culled, as Peirce puts it, from "our Tablet of memory," in which we store useful facts and memories "respecting the country of our birth or adoption."[30]

Each date or event is listed as if it were an itemized line in an account book wherein weather disasters are modified and balanced out by economic improvements and technological innovations. While there is no attempt to hold on to a strict continuity between weather conditions in Old England and New England, as Puritan settlers had done (in 1686, Increase Mather had noted that the Massachusetts climate "agrees well with the temper of our English bodies"; French settlers in New Orleans appraised the Mediterranean features of the Gulf climate), the character of the national weather is still chronicled in relation to European weather events, thereby historicizing its "natural" continuity with the Old World. Aside from his close attention to the historical growth of communications, crucial to the discursive mapping of a continental landmass like the US, what is most striking about Peirce's method is his attachment to facts and statistics. Like Thoreau's obsessive accounting of the costs of living on Walden Pond, Peirce's balance book of statistics establishes social and historical continuity for his record of the economy of nature. In doing so, he helps to construct in national terms what Daniel Boorstin was later to characterize as a "statistical community."[31]

In contrast to Peirce's amateur-gentlemanly endeavors, Espy's ambitious efforts were publicly tied to a professional concept of national weather, philanthropically conceived as a service for "the Farmer, the Mariner, and all Mankind." Indeed, these professional efforts formed the basis of an early federal meteorological agency, initiated as a service of the Signal Corps of the War Department, and helped along by Joseph Henry at the Smithsonian Institute, who centralized the service by developing a national network of observers reporting to Washington via telegraph. Established as the Weather Bureau in 1870 (administered by the War Department, and later under the civilian auspices of the Department of Agriculture (1891), and the Department of Commerce (1940)), the agency finally became the National Weather Service (NWS) in the 1970s, administered by the Environmental Science Services Administration and, today, by the National Oceanic and Atmospheric Administration. Each stage of its evolution has been marked by political factors, like that of wresting civilian status from the early militarized Bureau, and economic considerations—the shift to the Department of Commerce came at a time when service to aviation surpassed the service to agriculture.[32] While the information provided by the NWS today is available for next to nothing, the media, especially TV, spend millions on staffing outlays and simulation technologies in order to humanize this information and to present it as an entertainment segment on news broadcasts. But the NWS no longer functions as a centralized government monopoly. At least one private weather service—Accu-Weather, from Pennsylvania—runs an extensive and highly lucrative weather service for private enterprise, including many television stations. Accu-Weather's claim to greater accuracy was initially based on its close attention to European forecasting systems, which make use of computers that can produce more advanced simulations of weather movements than are available to the NWS. Joel Myers, president of Accu-Weather, claims that the NWS relies "more heavily on their own model than on the European model for nationalistic reasons" (*New York Times*, February 15, 1987). European superiority in this respect, often represented as a matter of grave national concern for the underfunded NWS, is also a major factor in the new transnational geopolitics of information, in which many sovereign states have limited access to information about their own countries that is gathered and held by multinational companies. In recent years, vast appropriations have been secured from Congress for a complete update of the nation's meteorological technologies.

Decentralization was considerably advanced during the Reagan eighties, when private forecasting services for all sorts of specialized activities flourished. Many private weather users can now receive images direct from satellite, without any centralized mediation or meteorological interpretation, while onboard weather-monitoring hardware has replaced centralized port systems, obviating NWS forecasts that are often outdated by the time they reach ships and planes. Decentralization and deregulation have not entirely displaced federal control over the interpretation and dissemination of national weather information, but it is clear that the concept of transnational weather has become a political and economic commodity in recent years. World weather has been a diplomatic issue, in principle, since the first meeting of the International Meteorological Organization in Vienna in 1873, while Washington's political interest today in the National Climate Program lags well behind its interest in the World Climate Program. On the one hand, the official recognition that weather is a global phenomenon means that acid rain, radioactive fallout, and global warming can be addressed as occurrences that do not respect national borders. On the other hand, the concomitant deterioration of sovereign national control over weather information has meant that decisions about its use and distribution can be controlled by, if not restricted to, private corporate empires or multinational cartels.

TAYLORIZE OR DIE

The efforts of Espy, Henry and others at the early Smithsonian to centralize the nation's weather by making use of the new telegraph technology looked forward to the progressively rational organization of our everyday life through advances in telecommunications systems. They also oversaw the emergence of an imagined national community, united, state by state, by a discourse that links "America" with the rhythms, crises and patterns of "nature"—in other words, a natural community, with natural borders, and with a natural, even providential destiny in the social world. In recent years, the coherent shape of that national picture has been altered to accommodate global concerns in ways that are still not politically fixed, let alone rooted in appeals to nature.

There is no question that the way we "see" the weather today is different from the way Espy and his contemporaries did. This is not simply because of linear "progress," associated with superior technology, advancements in the science of meteorology, or a less parochial understanding of the workings of the natural world. What stands between us is a dense social history in which "the weather" has been shaped and appropriated by various state and commercial interests. Meteorology, which only came into its own as a scientific discipline based on positivistic laws after the turn of the century, has been especially subject to market forces and military–industrial influence. In his book about the career of Joseph Bjerknes, the Swedish scientist associated with the influential Bergen school of meteorology, Robert Friedman provides a case-book example of this process by showing how Bjerknes sought prestige in the scientific world by legitimizing the new "science" of meteorology at a time when the fledgling aeronautical industry required accurate information about the atmosphere. The new meteorology proved a vital military asset during the Great War, and weather forecasting soon became a lucrative commercial service for air travel after the war. Bjerknes's "discovery" of air-masses moving in discontinuous "fronts" was therefore presented at a particularly opportune moment for transforming the discipline and indeed the whole practice of weather forecasting. Friedman shows that Bjerknes's new explanatory model of discontinuous fronts took its conceptual shape from a militaristic view of the northern hemisphere as a battleground upon which a warm equatorial current struggled against the cold polar current. Consequently, Bjerknes proposed a discursive picture of a single "battlefront" stretched around the hemisphere in a metaphoric extension of the kind of warfare that had been waged recently in Europe. Bjerknes's theory of the front formed the basis for a circumpolar weather service, with a school of disciples spreading the word throughout Europe and North America. With the authority of the theory established in scientific circles, it wasn't long before everyone began to "see" fronts. Friedman's story of the career of Bjerknes is a striking parable about the ways in which this scientific theory was advanced through power, authority, persuasion and responsiveness to commercial interests.

The existence of the "front" was, of course, a theoretical postulation, not something waiting empirically in nature to be discovered. To transform the accepted method of "seeing" and "reading" the weather, Bjerknes had to invent a weather semiology that would be more stochastic and less static than the old mechanistic method of observing.[33] A system of signs was needed by which observers could interpret the existence of these discontinuous fronts before they arrived. Friedman describes how Bjerknes sought support for his theory on the margins of science, and on the geographic margins of the nation:

> Using clouds as signifiers of atmospheric processes occurred to Bjerknes during the setting up of the Bergen forecasting service. While traveling up and down the coast in 1918, to the outermost reefs and to the remote ends of the fjords he acquired new insights into weather prediction. From the farmers, lighthouse keepers, fishermen, and sailors who were to be observers for the forecasting service, he learned a rich folklore of weather prognostics. Interwoven with ancient and superstitious beliefs were useful predictive signs. To help compensate for the missing data from the west, he realized, a system of signs might be devised that, when interpreted correctly, might provide clues to the weather systems approaching from beyond the North Sea horizon.[34]

Here, surely, are all the marks of a heroic narrative, wherein the intrepid scientist redeems the latent half-truths of popular consciousness by converting followers of folklore and superstition into amateur disciples of science. To provide confirmation for his theories, Bjerknes established a set of forecasting instructions for these untrained observers. The instructions were designed, of course, to fit the theory, to see nothing but evidence of discontinuous fronts in the clouds. The existence of secondary fronts, for example, long suppressed by the Bergen theory of a single polar front, was excluded, and rendered impossible by the forecasting system constructed by Bjerknes. If, in Bjerknes's transformation of the sailors' weather folklore and prognostication into a systematic method of prediction, we see nothing but a shining example of enlightenment progress that we have happily inherited, then we see only half the story. The other half tells us about the social factors that helped to shape that transformation in ways that had more to do with politics, commerce and prestige than with inevitable advances in science.

A similar caveat could be offered in considering the difference between the weather in Espy's golden age of rainmaking on the frontier, and the weather in today's globally warming world. In the course of that history, too long to recount in detail here, it could be argued that a number of large-scale cultural transformations have taken place: the "gay science" of weather folklore and divination by consulting plant and animal behavior has been replaced by our new "unhappy" relation to the meteorologists' jargon-laden language of forecasting; the everyday cult of "experience" has been replaced by the professionalist ethos of "expertise" in responding to weather semiology; the rich array of weather perceptions—how we listen to the weather, see and feel it coming and going—has been replaced by a spectrum of "weather sensitivity"—the vulnerability of our bodies to the minutiae of weather phenomena (a certain division of mental and manual labor remains in place—working people have weather-sensitive bodies, while intellectuals have weather-sensitive minds); a code of early-warning systems for the defense and survival of local, agrarian communities has been replaced by a massively institutionalized anxiety about the defense of national shorelines; weather-as-news on the scale of local tragedy has been replaced by weather as commentary on regional, national and international events, especially weather disasters high on the scale of geopolitical tragedy; the quackery that once accompanied the culture of almanacks and patent cures for weather-induced ailments has been replaced by the almost complete commodification of bodily maintenance in the face of year-round weather threats and assaults; "looking out the window" has been replaced by the perspective of a geo-stationary satellite; and our mental or cognitive maps of the psycho-geographical environment have been replaced by the objectively simulated representation of an environment under the influence of weather, other people's weather that is only temporarily our own.

This list could go on and on, but its structure already suggests an argument that I don't really want to make. Such a list of changes, narrated in such a manner, lends itself to an already well-known story, told by the left and right alike, about the increasing technorationalization of the object world and the concomitant increase in the methods of social control and production of passive, commodifiable personalities for us all. While I don't intend to take issue with this narrative (and am otherwise quite sympathetic to many of its political assumptions), I do not think that the story it tells about the exercise of cultural power is an altogether adequate one. This story, governed by a nostalgic narrative of decline and disenchantment, locates "value" in the relative coherence of folk, communitarian *experience*, increasingly eroded and alienated by modern technologies. A counter-narrative would simply assert the primacy of knowledge and information (categories no more value-free or any less mediated than that of "experience") over ignorance and superstition. It would point out that pre-technological "proximity" to nature, far from guaranteeing anything that resembles natural or social equality, was one of the primary means of maintaining a repressive social hierarchy. Indeed, only in recent years has proximity to nature, "untouched" by mankind, acquired a positive value in our culture—a turnabout from centuries of association with barbarism, demonism, and worse.

Raymond Williams has pointed out that the dominant ideas about nature contain "an extraordinary amount of human history." In the modern period of Western culture, Nature has been personified as, respectively, God's (medieval) deputy or minister, an absolute (Renaissance) monarch, an (Enlightenment) constitutional lawyer, a (Darwinian) selective breeder, and a (laissez-faire) free marketeer, among others.[35] It could easily be argued that none of those roles guaranteed the general population anything but a passive, dominated fate, subject to the brutal "laws" of "natural" scarcity and necessity as created or interpreted by ruling interests. There are few today who would dispute that theological knowledge about the providential visitation of storms upon a backsliding populace is a less empowering explanation of the weather than scientific knowledge about the disastrous contribution of industrial processes to atmospheric degradation. But this is a false comparison, and offers no real explanation of the different sets of power relations that could be mobilized around each of these claims. We need to consider each such claim in its own social and historical context, taking the time to describe the interests it served and the local politics to which it gave rise at a particular time and in a particular place. For every example of a repressive pre-technological social hierarchy there is a counterexample of non-hierarchical social existence at a low level of technological development.

Dominant ideas, whether in science or theology, are legitimized and enforced at any time by presenting them as part of a natural order of things. Ever since science, in the Enlightenment narrative of progress, was posed *against* an intractably stable order of nature, the maintenance of cultural and economic power has rested upon a dialectic of change and constancy, innovation and stability, progress and conservation. Consequently, the game of winning general consent for ideas in the history of modern capitalism has been waged on a balanced terrain of contradictions, where narratives about traditional values are played off against narratives of progress. Something always has to be sacrificed for progress to be made, and yet progress is only sanctioned if what it displaces is preserved in some other region of social life. It is in the context of this play of values that any history of modern weather can best demonstrate its role in the national culture. For the way in which we talk about weather patterns of change and repetition is fundamentally linked to the dialectic of change and constancy that lies at the heart of a developed capitalist culture. To focus on the ever-modified shape of that dialectic from moment to moment is to reject the explanatory power of ultimate linear narratives about progress, whether those of progressive domination or progressive emancipation.

In this respect, the recent hysteria about global climate change—the transformation of a natural phenomenal order that was hitherto assumed to be constant—can be seen as a continuation of historically regional and national anxieties. The history of US weather is full of examples. Take the annus mirabilis of 1816—the famous "year without a summer" in which frosts and snows persisted all summer as far south as Pennsylvania and New Jersey. Crops failed, famine followed, the shift in New England from an agrarian base to industrial manufacturing was initiated, and the great migration to the Middle West was begun. The popular perception of these events, reflected in Jefferson's suspicions about climate change cited earlier, was that the weather had permanently altered, and thus that years without summers had become the rule and not the exception. Or consider the coincidence of severe economic depression in the 1930s with drought-like conditions in the Dust Bowl midsection of the country, from Mexico to Canada, and from the Great Lakes to the Sierras. The Dust Bowl disaster proved that semi-arid lands in the West could not support the same farming methods used in the more humid East, precipitating a similar migration west to California. Again, the popular perception was that the weather had changed permanently—for the worse—a perception that meteorologists at the NWS were hard put to combat. A third example, somewhat more impressionistic, is the recent generational memory that the US, in the period from the early fifties to the early seventies, enjoyed a prolonged bout of benign weather, discursively linked, of course, to the climate of national health: the period of political consensus, affluence, unimpeded economic growth, and consumer benefits for all. In contrast, ever since the OPEC oil embargo, the rise of multinational capitalism, and the onset of ecological anxieties, we have seen the breakup of that national weather consensus and the growth of a disaster culture, fed by the new global fears about melting ice-caps and the like, which is now presumed to be a fixed feature of the climatic future.

Instances of prolonged meteorological abnormality expose popular and official anxieties about the economy of change and constancy that regulates our everyday lives. Historical weather events, no matter how singular or prolonged, are remembered as material instances of radical abnormality long after contemporaneous political or social events and upheavals have faded from the popular memory. Famous blizzards, droughts, tornadoes, and hurricanes punctuate regional and national history with a social meaning that often far surpasses the resonance of the political events with which they coincided or were causally associated. The great Texas drought of 1885–87, for example, was such a significant moment in the history of white "conquest" of the state that Texans' everyday reference to historical events was qualified by the markers "before the drought" and "after the drought" far into the next century.[36] More recently, the summer of 1988 has played a similar role as a memory marker for popular ecological consciousness. The long-standing endurance of these popular memories is no doubt partly due to the perception that, unlike remote political events, natural hardships, at least in principle, affect everyone almost equally. But it is surely also an example of the way in which change, even the most extreme instances of change, is habitually naturalized, or at least presented as the result of natural, not human or social, causes.

Changes in the weather from day to day are our most palpable contact with the phenomenon of change, and so it is no surprise that they often come to be associated with patterns of social change. When the weather does not change for weeks, anxiety about stagnation sets in, just as business economists worry about an inactive economy in need of stimulation. Under drought-like conditions, the weather affects the economy directly because it withdraws its services to industry that nature provides (or used to provide) for free. On the other hand, when the weather calls particular attention to its mutability, we call it "changeable weather" and reserve for it those terms that are patriarchally associated with the feminine; March, for example, is "fickle," July is "sultry," and the practice of feminine naming of hurricanes was, until recently, reserved for the most unpredictably violent of tropical storms.

In national weather cultures like that of Britain, weather mutability and the climate of endurance it inspires are so deeply pervasive that the weather is often described in the ambivalent terms reserved in the popular consciousness for a nationalized institution. In the wake of Mrs Thatcher's victory in the general election of 1987, for example, the debate about the intransigency of nationalized public services was stepped up by conservatives. William Plowden, Director General of the Royal Institute of Public Administration, tried to characterize a now outdated consensus mood in the following way: "It was widely agreed that public services, though in principle admirable, were inherently conservative, extravagant in their use of resources, labyrinthine in structure, liable to give their customers not quite what they wanted, and impervious to criticism. Little more could be done about all this than about the weather" (*The Independent*, June 24, 1987). In the United States, the weather may be national but it is not nationalized; we submit to it not as we would submit to the State—like it or lump it—but rather as we would submit to the "invisible hand" of the market and its so-called free and natural laws. In this respect, everyday change is necessary, no matter how random and contingent it may appear, if the appearance of the system's stability is to be guaranteed; as weather citizens, we are prompted to greet and encourage our investment in change. Things have to change for the system to reproduce itself; this is the fundamental principle of a market economy, and who would expect anything but capitalist weather in an economy where the ultimate reward for a lifetime's service to business is to go into weather retirement in one of the Sun States or the SunBelt. This, then, is the built-in incentive of the capitalist weather system. No other nation is geographically equipped to promise and deliver such benign rewards for such a large population, while maintaining its Puritan work ethic by invoking the rigorous extremes of seasonal changes in the North. The contrast between the climates of the two non-contiguous states, Alaska—frozen hell of labor and oil production—and Hawaii—balmy paradise of recreation and vacation—expresses the symbolic extremes of the pre-retirement weather system. In contrast to an imperial weather system like that of the old British Empire, covering territories where "the sun never sets" and thus marked as a timeless, boundless, and changeless (political) condition, the American weather system, like its earliest system of industrial mass production in the nineteenth century, is a dynamic mechanism of interchangeable parts.

Consider Mark Twain's famous characterization of New England weather, in which this economy of "production" can be seen, reproduced on a regional basis:

> I reverently believe that the maker who makes us all makes everything in New England but the weather. I don't know who makes that, but I think it must be raw apprentices in the weather-clerk's factory who experiment and learn how, in New England; for board and clothes, and then are promoted to make weather for countries that require a good article, and will take their custom elsewhere if they don't get it. There is a sumptuous variety about the New England weather that compels the stranger's admiration—and regret. The weather is always doing something there; always attending strictly to business; always getting up new designs and trying them on the people to see how they will go. But it gets through more business in spring than in any other season. In the spring I have counted one hundred and and thirty-six different kinds of weather inside of four-and-twenty hours. It was I that made the fame and fortune of that man that had the marvellous collection of weather on exhibition at the Centennial, that so astounded the foreigners. He was going to travel all over the world and get specimens from all the climes. I said, "Don't you do it; you come to New England on a favorable spring day." I told him what we could do in the way of style, variety and quantity. Well, he came and he made his collection in four days. As to variety, why, he confessed that he got hundreds of kinds of weather that he never heard of before. And as to quantity—well, after he had picked out and discarded all that was blemished in any way, he not only had weather

> enough, but weather to spare; weather to hire out; weather to sell; to deposit, weather to invest, weather to give to the poor.[37]

It would be impossible to better Twain's canny feel for the local business pride and acumen of the well-to-do members of the New England Society, to whom he addressed this official dinner toast. His is a character act that does wry justice to the culture of the huckster and the genteel industrialist alike, gently demolishing the boundary of "good taste" between the two that his audience would no doubt have preferred to maintain. In juxtaposing the traditional system of craftsmen/apprentice labor with the more mercenary exploits of the roving showman–profiteer, Twain gives us a concise picture of the changing styles of business culture in the newly industrializing mid-nineteenth-century North-East. The profiteer is still posed as an outsider, a "stranger," but his methods and ways bring him "fame and fortune" by exploiting the local product, labor, and natural resources of the region. The "dazzling uncertainty" afforded by the sheer diversity of these natural resources is a testament to the cornucopia of opportunities available to all would-be profiteers. What Twain goes on to call the "inhuman perversity of the New England weather" is here transformed into a marvelous investment opportunity. Nature's variability, a source of great discomfort to the "patient and forebearing" New Englanders, becomes a commodity of the highest promise, a second nature whose inconstancy will be transformed into a source of constant revenue.[38]

If Twain's story can be read as a parable of the exploitation of natural resources, it also describes the reverse. Natural diversity is invoked to describe a system of variable production as if it were coherent: commodity production as a natural process of risks and opportunities. Twain's New England weather system would be extended in the following century to its fully national dimensions where what he called the "size" of the region's weather still retained its cornucopian promise. A century later, Archibald MacLeish, in a celebratory Popular Front mood, would invoke this promise with none of Twain's deft skepticism: Americans "had the luck to be born on a continent where the heat was hotter, the cold was colder and the mornings were more like mornings than anywhere else on earth."

The ideological maintenance of the national weather system depends, as I have suggested and as Twain's speech illustrates, upon its naturalization of social causes and effects. If extreme weather conditions can cause suffering, then people's suffering can always be explained (away) in turn by weather conditions; the foreclosure of a family farm can then be seen as an inevitable component of a natural cycle. Inequalities are evenly distributed throughout the country according to the current weather configuration of highs and lows; some like it hot, some like it cold, some times are good, some are bad, wherever there are winners, there will be losers elsewhere. Rewards and punishments are disbursed in a cost–benefit analysis that balances out across the breadth of the nation. This weather system of credits and debits refers as much to a political model—the US constitutional system of checks and balances—as to an ideology—the holistic maxim that "Mother Nature will balance everything out." Such appeals make it seem as natural as the weather that some people must experience hardship so that others may enjoy abundance.

Similarly, on the global scale, regional and hemispheric inequalities can be explained according to geography and not history. Nature intended the distribution of weather wealth to be this way, and if the developed countries happened to have enjoyed the benefits of a temperate climate, then that is because human industry naturally favors such a climate. Famine in underdeveloped cultures comes to be seen as a determined effect of the climate rather than a result of agricultural decisions (to grow monocultural crops for export, like cotton and coffee and bananas, instead of grain and other local food crops) imposed upon such countries by the World Bank. Modern "civilization" belongs in the northern hemisphere; nature has determined that it cannot flourish elsewhere. A climatological map is used to *explain*—and increasingly, in the age of global warming, to *enforce*—unequal development, where a map of colonial exploitation, influence, and dependency would give us a much more accurate picture of the histories behind these inequalities.

PUTTING THE MAP IN MOTION

Until now, I have been stressing the way in which the weather is used to naturalize the social: appeals to weather and climate help to explain a set of contingent events as if it were a condition of nature and not a condition of society. In an imagined national community, the weather plays nature to the culture of our social and political life, while it provides a broadly understood language for helping us to make sense of that social and political life. This is the primary function of the long weather segment in local news broadcasting, where the weather acts as an insider trader, playfully mediating between news and sports. Here, the point is often to make weather out of the news events by underlining the links between hardship and good fortune, between nation and region, between the unalterable nature of political logic and its often contingent, local effects. A similar function applies to the economic climate. Statistical estimations of the day's weather are often presented before or after an announcement of the stock prices of local companies. This function came into its own during the Wall Street "crash" of 1987 when the specter of falling stock prices was a newscaster's favorite segue to the forecast (in the eastern half of the nation) of falling temperatures. In reports about the situation in Britain, the "storm" that hit the London Stock Exchange was figuratively matched with the hurricane-force storm that had hit southern England a day or two before; both were cast as terrible acts of nature that had arbitrary effects upon, but no explicable causes in, the social world.

With the advent of the Weather Channel in 1982, a 24-hour cable service broadcast by Landmark Communications in Atlanta, the institution of weather forecasting took on a decidedly new national "character" and a new purpose—to make news out of the weather, rather than to make weather out of the news. The channel's commercial success is driven by what its managers describe as its "continuing goal of becoming the nation's primary source of weather information." Programming on the Weather Channel consists of a ceaseless flow of different narrative segments and features strung together in a highly organized schedule. Fast-paced regional weather information in the morning (moving across the time zones—eastern, central and western—as the morning progresses), is succeeded by more relaxed entertainment and lifestyle features—"Weather and You"—in the afternoon, followed by a faster series of updates as the evening progresses, to capture the "zapper" audience at the top and bottom of entertainment shows on network TV. Local forecasts are broadcast twelve times per hour in the morning, eight in the afternoon, and ten in the evening. Regular features at least once every hour throughout the day include Aviation Weather, Boat & Beach Reports (in the summer)/Skiers Forecast (in the winter), Business Travellers' Report, Fishing Forecast, Michelin Drivers' Report, Schoolday Forecast, Tropical Update/Winter Storm Update, and the Weekend Outlook/Look Ahead. Even the average viewer (10 million households per week tune in) who habitually watches a few minutes each day recognizes that he or she has hooked into a highly developed universe of discourse about the weather. The Weather Channel is self-contained: "the programming works," claim its managers, "because weather is a universal subject, and everyone has some kind of weather need." Viewers find a constantly varied presentation of scientific information, friendly advice, and spontaneous philosophy; accessible, concrete displays of otherwise abstract weather events; immediate, almost indexical, relations between commercial sponsorship and weather features after the fashion of the direct sponsorship favored on radio and early television (features on cold snaps are sponsored by Thermalite Thermal Wear; allergy and health features by Afrin Nasal Spray; international weather by AT&T); and a corresponding dynamic between weather threats or crises, both national and local, and pragmatic solutions to be found in the commercial world's advertised products. Indeed, much of the channel's commercial success has been due to its pioneering use of a cooperative advertising program that allows local dealers to tag national network spots. This "creative advertising" allows a filmed national commercial to include the insertion of local text tags, identifying only the dealer(s) covering each area or neighborhood. The program is made possible by the same technology, called the Weather Star, that allows local weather broadcasts to be inserted into the national flow of the cable transmission. With this technology, the local joins seamlessly with the national; the local economy of the national weather system is delivered by the same logic with which a national corporation hooks up with a local affiliate.

For weather addicts and enthusiasts like myself, the world of the Weather Channel is a coherent fiction, with its own generic narratives and its own formal agents, like the jet stream, developing fronts, anticyclones, troughs of low pressure, and so on. The interpretive commentaries of the On-Camera Meteorologists (OCMs) are a pleasurable invitation to travel through this fictional world: "let's put the map into motion . . . well, the weather is making a quiet start across the country, but as daytime heating occurs, we'll see developments"; "not much to talk about in the West, but there's an evolving storm over the Rockies, and an interesting batch of rain over the Great Lakes"; "another hot day on tap in the southwest, but there's a cool spot up there in Maine, and let's see what's causing all this rain in the Ohio Valley"; "a dome of high pressure over the Great Basin is responsible for this oppressively hot temperature regime in the Plains States," while "a trough of low pressure down in the Gulf is acting like a factory, generating storms throughout the southeast." The prevailing discourse is one of national checks and balances: "temperatures are building in the East, with compensation in terms of Western cooling"; "it's very quiet in the Caribbean, but up here in the Pacific northwest, the storms are lined up, back to back." And the international forecasts are shaped by a national world-view that sees continental blocs as comparatively equal to the meteorological unity of the US. "If you're wondering what the weather's doing overseas . . . here's the international forecast" (usually only Europe, but more recently the Persian Gulf region, where weather-for-war was a vital military and commodity interest) where the same "scale" and "range" of coverage is applied to the length and breadth of a continent with highly distinctive nation-states: "not much in the way of precipitation, except up in Finland"; "the heat is on in Greece and Italy, while a ridge of low pressure extends all the way from Western Russia to Western Ireland." As a transplanted European, I know that such a continental forecast would make little or no practical sense to the national weather citizen of any European country. Rather, the forecast is tailored to the Brobdignagian perspective of the American weather citizen, who is addressed as a weather tourist rather than as a real, or potential, travel tourist and who is asked to extend his or her macro-geographical perspective of the United States to that of Europe. Countries sharing the North American continent receive attention only when "their" weather is seen to be affecting "ours," or when they "share" bioregions like the Great Lakes, the high plains, the Pacific Northwest, or the Mexican Gulf. On the weather map, Cuba, for the most part, still does not "exist"; it does not influence US weather, since the US has no influence over it.

It is clear that this fictional world of the Weather Channel is selectively shaped by a social and political mapping of the world as much as it is determined by the atmospheric map of shifting fronts and air-masses. The same could be said of its commercially inflected appeals to everyday life. Here, the Weather Channel fully embodies the specialized division of labor between meteorological services that support weather aimed at consumers/ households—who use weather information for welfare, safety, health, and planning leisure activities—and weather aimed at producers/companies—who use it for short-term profit. Producer weather is for agriculture, forestry, marine and aviation needs, and for industry. On the Weather Channel, there is no weather-for-work, only for leisure and consumer time. Thus while the channel regularly features gardening segments—frost freeze maps, planting calendars, and "lawn and garden" reports—no attention is given to the exigencies of agricultural production, except, of course, in the case of extreme or emergency conditions of drought, flooding, hurricanes, or unseasonable frosts which may directly affect consumer markets. In the Weather Channel's world, people do everything but work; weather affects how they "drive" to work, and travel to work if they are "business travellers," but it has no bearing on their actual work environments, which are assumed to be immune to the weather. The absence of "producer" weather makes this a postindustrial world, where any evidence of labor is located in the service sector ads, on the business class air routes, in the household, for women, and in DIY activities for men at home. Consumer activities are limited to the range of average upkeep. Fashion discourse, similarly, is restricted to thermal underwear, or clothing the children for school, or keeping the elderly well wrapped up. Eco-conscious discourse is restricted to water and energy conservation tips: among the many Weather Channel maps, there are no maps of acid rain damage, deforestation, oil spill concentrations, toxic dump locations, or downwind nuclear zones. In the absence of these politically complex health and safety hazards, the responsible weather citizen's rights are only threatened with natural and not social erosion. So too, the channel's multiple address to individual, (his) family, and nation is pluralist in principle but speaks primarily to the citizen identity of a white male property-owner. Ideal Weather Channel "citizens" are assumed to be comfortably off, white-collar, with cars, boats, vacation options, families, and gardens and homes that require extensive upkeep.

While the Weather Channel shares many of the structural formats of the respective TV products of the news industry, the culture industry and the leisure industry, it services each of these industries in its own right. In fact, it is the first fully developed product of a commercial service industry we have seen on national television. The Weather Channel no longer "produces" the weather, and layers of discourse about the weather, as a service designed to meet certain perceived needs, but rather as an entire outlook on everyday life, even a "conception of the world" as determined and explained by the weather.

"You need us for everything you do," the slogan that most frequently announces a station identification on the Weather Channel, also announces the complete "weatherization" of everyday life. The weather-effect is presented as omnipresent in our lives, from the dust that settles around us as we sleep to the pockets of air stagnation we pass through on our way to work, to the larger psychopathologies associated with storm behavior. To remind us of our exact physical location in this spectrum of influence, the Weather Channel offers an almost inexhaustible series of national maps (aside from general elections, weather forecasting is the only time most of us see a national map): fishing maps, business travel maps, picnic maps, indoor relative humidity maps, outdoor relative humidity maps, tanning maps, allergy maps, humidity maps, the ominously named "aches and pains index," influenza maps, precipitation maps, radar maps, storm history maps, windy travel maps, and many more—each charting in detail the geographical distribution of daily weather effects on our bodies, and each sponsored in turn by the manufacturer of an appropriate product.

But this complete weatherproofing of every commodifiable moment of our lives suggests more than a relation of dependency on the "weather fix," i.e. specific consumer solutions for specific weather-related problems. Rather, it suggests an advanced form of weather-consciousness. We might be tempted to call this "false weather-consciousness," but I think there are good reasons for not doing so—even if, at times in this chapter, "weather" could be read as substitutable for "ideology."[39] In expanding the definition of weather to include all of the ways, forms and contexts through which our body responds to and is constructed by discourse about the environment, the success of the Weather Channel, the success of the full-color weather page in the national newspaper *USA Today*, and the general explosion of interest in the weather throughout the eighties, speaks to new configurations of power that increasingly devolve upon the everyday body.

"Weather-sensitivity" has become a pressing new feature of our concerns about social adaptation. As the author of one popular guide puts it, "Learning what the weather can do to your body and emotions is the first step toward coping with weather sensitivity and functioning at your full potential. Understanding the weather sensitivity of others can help you on the job, in your love life and with your own family."[40] Executives at the Weather Channel attribute the new weather-consciousness to the needs generated by increased mobility of the general population. Some commentators point more directly to the new interest in environmental issues, while others see the exploitation value of disaster weather as entertainment. Hooking both of these together, newspapers like the *San Francisco Chronicle* and the *Chicago Tribune* now run a regular global weather feature called "earth week," "a diary of the planet" that pinpoints fearsome floods, tornadoes, volcanoes, earthquakes, space exploration, ecological problem areas, and the like.

What is less clear is the shape of the new biopolitics that is emerging from advanced weather-consciousness. If we are to extend Michel Foucault's speculations about the exercise of modern power as disciplinary knowledge in the form of scrutiny or surveillance about the body, then we would say that it is the weather-sensitive body rather than the weather itself that is the visible object of all of this new knowledge. Anyone who watches the Weather Channel for the first time is surely left with an altered perceptual sense of how his or her body functions in a number of environments. So too, the barrage of statistics, from the present, the past and the future, that accompanies the weather forecast creates a field of knowledge and constraint for the body. Discourse that situates the current weather in relation to a history of weather statistics functions as a way of normalizing our physical life, regulating its mean or average behavior in relation to an archive of temperature records. Abnormalities like record highs or record lows are part of the regulatory field of differences that locate our current degree of deviation from a norm of environmental behavior for which we are then made to feel responsible in some way. Statistics about the mean, norm or average belie the fact that there is no such thing as "normal" weather, let alone a "normal" climate; these average figures play the role of normalization for us. We, no more than our ancestors who may have been subject to the same climate, did not *create* these statistics, but they are now part of our responsibility, both to the local community and to the national community, with whom we share the weather in different ways. Weather citizenship now comes with that quantified sense of historical and geographical responsibility. Knowledge about global warming will add a new layer of responsibility, a new level of concern whenever and wherever the weather exceeds or deviates from the average, statistical norms, while the disaster culture to which "global warming" belongs will continue to function as an effective way of symbolically managing the behavior of mass populations.

But these new forms of normalization and regulation, however constraining, are also accompanied by new and emergent forms of politics, waged with scientific knowledge and information, around bodily and environmental concerns. The "weather" has become a politicized environment precisely because it is an effective medium for linking biopolitics to large-scale forces in the social and physical world. The global warming debate, as I have argued, may yet become a powerful vehicle for the controlled re-formation of citizen subjectivity, but it also brings into play the opportunity to wage other kinds of politics, with other kinds of rights, liberties, and responsibilities than those attendant upon a laissez-faire view of the world.

It would be a mistake, then, to conclude that what I have called "advanced weather consciousness" is simply part of an extension of social control over our everyday lives, and that the Weather Channel, say, is a highly engineered commercial exploitation of that process. It is likely, for example, that viewers use the Weather Channel for all sorts of different purposes, in ways that cannot be wholly explained or anticipated by a "textual" analysis of its own mode of address or presentation, and its construction of weather-vulnerable individuals who belong to a coherent, national community. We cannot afford to see the Weather Channel's organization of these features as a coherent TV discourse, inevitably producing a similarly coherent TV viewer, inert, socially isolated, and incapable of re-formation. TV is never, or rarely, watched as such. We know that television today is the "great reproducer" of existing social relations, but in the overall course of its production of social knowledge it is many other things as well. To merely describe the reproductive capacities of television would be to defer to the kind of history I have tried to avoid, a history that equates advances in technological rationalization with increases in social control and domination, and thus a mode of information that inevitably produces blindly patriotic, weather-fearing citizens.

A similar caveat might apply to arguments about nationalism that have surfaced in the course of this chapter. To use Benedict Anderson's term, a "nation" is an "imagined community," the construction of a "deep horizontal comradeship in anonymity," bound together by a created sense of common history—or, just as relevant to weather scholarship, a common geography.[41] In other words, the nation is not a false or artificial community, least of all today, when it still stands as one of the most universally legitimate values in political life. But to *only* describe its creation or historical formation is always to take the side of the creator. And to insist, in our histories, on the overriding importance of the construction of nationalism is, more often than not, to over-subscribe to the power of its effects. Rather than stress alternative narratives of US nationalism, which emphasize its republican, revolutionary origins, we cannot afford to assume that national identity can be fixed or given in ways that are simply more attractive to the left. We ought to be able to emphasize the limits and failures of national identity just as insistently as we describe the successes of its construction. In many countries, where nationalism has summoned up more collective hates, desires, and deathwishes than class struggle has ever done, the "effects" of nationalism have sometimes been politically progressive. In the US, where the narratives of nationalism have traditionally been a conservative preserve, this is less often the case. Unlike those countries where, historically and even now, nationalism is associated with populist revolt, US nationalism emerged out of a fear of lower-class mobilization; a fear evoked originally by the prospect of extending tolerance, if not freedoms, to seditious slaves. In this respect, the structural history of US nationalism has been one of containment rather than one of self-determination. This no doubt accounts for its relative weakness, and, consequently, for the paranoid-aggressive terms in which it is periodically affirmed by the right wing. Indeed, as Anderson points out, the affective bonds created by the earliest formation of nationalism were hardly so strong that they could not prevent a war of secession a century after the Declaration of Independence. (Nowhere in the Constitution can the term "national" be found, incidentally.) A century later, the short-lived moment of the US nation-state was virtually over.

The upsurge in nationalistic fervor of the Reagan–Bush years may have been a sorry symptom of the decline of the nation-state, nothing more than a slick cover for speculators to plunder and pillage in the name of national pride. The pageantry of red, white and blue was an especially hollow response to citizens of color whose rights and aspirations were clearly at odds with the dominant version of American nationalism. As 1992, the quincentenary year of Columbus's voyages, approaches, let us hope that there will be ample public opportunities to contest the official definition of "America" and to recount its imperialist history as one nation-state among others in the Americas.

On the other hand, the successes of Reaganist and Thatcherist nationalism clearly exploited public indifference to any conception of global identity, even of the genre favored by the media in its sustained enthusiasm for McLuhan's picture of the global village. Eco-idealists are speaking of entering an age of global democracy, where the environmental rights of all global citizens will be more important than the interests of individual sovereign states with their restricted conception of citizenship. In view of the uneven and incomplete picture of social rights that is apparent in most existing nation-states, this sunny vision seems premature. More to the point, perhaps, it is a vision as dear to the heart of multinational corporate philosophy as McLuhan's globalism was over twenty years ago. But there is nothing to be gained from ceding the globalist ground in advance. Environmental rights, in particular, will be a necessary feature of the kind of globalism that might yet be truly internationalist.

POSTSCRIPT: BEYOND THE NATION'S SERVICE

The 1990 Weather Guide Calendar, published by the Weather Channel, carried a side-feature about the national network of amateur observers, active at over 10,000 locations across the nation, who maintain the volunteer service established over a hundred years ago at the Smithsonian. One such observer, Rudy Villareal, is quoted under the heading, "Why Do

They Do It?": "Being a Cooperative Weather Observer (gratis) is my way of paying the country for all the good and riches that I have received as an American citizen."[42] His remark, of which James Pollard Espy would have been proud, is intended to show the degree to which the amateur spirit is still infused with a sense of the economy of national identity. Clearly, this is the Weather Channel's idea of a responsible weather citizen. But a closer look at the culture of amateur weather observers (Why do they do it?) revealed a much wider range of motives and interests. The rough sampling of volunteer observers to whom I talked in upstate New York had much more to say about local community service than any sense of national, or even patriotic, duty, even when asked to comment specifically on the "national" angle of their voluntary activities. The stories they told about these activities were very personal histories of their respective interests in weather observing: for some, regular observations were a way of organizing daily life; for most it was related to their occupations in some way, as farmers, teachers, or in environmentally oriented jobs; for others, it was an extension of their enthusiasm for "nature"; one or two even confessed that it was a way of getting free weather technology from the government. Many expressed a skeptical attitude toward the consensus opinions of experts. When asked to comment on the "effects" of global warming, almost all referred to their own experience and memories of recent weather trends rather than to the data they collected for the NWS.

Most of the "official" weather observers I talked to were relatively isolated in their activities. Occasionally, they would get together with an acquaintance who shared their interest in order to talk about the weather, but none of them had confessed to having any social contact with the other official observers in the region. By contrast, socializing runs high in the regular meetings of many of the chapters and organizations affiliated with the Association of American Weather Observers. The Association's monthly magazine, *American Weather Observer*, is a lively record of the enthusiasm of weather fans for sharing weather stories and information. It features columns about historical weather, sensational features about storm-chasing in Tornado Alley (Oklahoma–Texas Panhandle–Missouri), short submissions from readers about weather anomalies, purple passages about awesome weather sightings, and records of observational data from around the country.

It is likely that any properly ethnographic study of amateur weather culture (beyond the scope of my research) would uncover a range of motives much more diverse—and even contradictory—than those suggested in the single quotation selected by the Weather Channel to represent the voluntary work of the amateur observer; motives that fall short of the demands of national service, duty, and responsibility; motives so attached to the local, regional, and everyday that they are even opposed to the dominant idea of national service, which puts the "nation's interests" above all else. It is likely that these motives would range from the articulate desire for an amateur community that does the work of professional experts in its own nonbureaucratic way, to the realm of more personal, liberatory desires. As one respondent to the *American Weather Observer*'s readers questionnaire put it: "the AAWO has unshackled my repressed desire to express myself regarding weather and associated activities."[43]

While few of the observers I talked to had a well-informed understanding of the theories and data that occupy the experts on global warming, all of them spoke with enthusiasm about related environmental issues. Republicans in their number openly lamented the Reagan–Bush policies, and, while maintaining their general opposition to "government regulation," all agreed on the need for strong environmental regulation across the globe. One or two confessed to a more theological acceptance of planetary behavior; global warming was probably God's will. While all agreed that the anxieties about global warming had probably prompted people to think more about the supranationalist nature of the problems, few thought that ordinary people had access to the political resources to do anything about it.[44]

These stories offer a brief and messy picture of the uneven conditions for any emerging sense of "global citizenship," although they can be read as supporting the thesis that "global warming" presents a pathway toward such an identity. Globalism is already a well advanced ideology in the highest political and corporate circles, displacing the sovereign powers of individual states and the internationalist blocs that have made world history until now. In the geopolitical arena, globalism has begun to masquerade as the New World Order, a front, at least in the Gulf War, for a new first world alliance of neocolonial interests. In spite of the media picture for a world "united" behind the Western war campaign in the

Gulf, there is no doubt that globalist sentiment and globalist awareness is still thin on the ground among the majority of the population. Let us assume that the widespread anxiety about global warming is about to bring us across one of the first hurdles. Before the race for globalism goes any further, we will have to do our best to ensure that it will not be a short dash, favoring the most highly trained and the most lavishly sponsored, but rather that the rules of the event, which must be open to all participants, have yet to be decided.

NOTES

INTRODUCTION

1. Murray Bookchin, *Post-Scarcity Anarchism* (Berkeley: Ramparts Press, 1971); *Toward an Ecological Society* (Montreal: Black Rose Books, 1980); *The Ecology of Freedom* (Palo Alto: Chesire Books, 1982); *Remaking Society: Pathways Toward a Green Future* (Boston: South End, 1990).
2. Christopher Lasch, *The True and Only Heaven: Progress and its Critics* (New York: Norton, 1991), especially pp. 22–4.
3. C.P. Snow, *The Two Cultures and the Scientific Revolution* (Cambridge: Cambridge University Press, 1959), p. 11.
4. Stanley Aronowitz, *Science as Power: Discourse and Ideology in Modern Science* (Minneapolis: University of Minnesota Press, 1988).
5. Donna Haraway, *Simians, Cyborgs, and Women: The Reinvention of Nature* (New York: Routledge, 1991), p. 192.
6. For a more exhaustive analysis of this recommendation, see Sandra Harding, *The Science Question in Feminism* (Ithaca: Cornell University Press, 1986).

CHAPTER 1

1. Paul Brodeur, *The Zapping of America* (New York: Norton, 1977). In his most recent book, *Currents of Death: Power Lines, Computer Terminals, and the Attempt to Cover Up Their Threat to Your Health* (New York: Simon & Schuster, 1989), Brodeur provides an exhaustive history of the attempts of individual researchers—Nancy Wertheimer, Robert Becker, Andrew Marino, Ross Adey, Karel Marha, Carl Blackman and others—to expose the dangers of 60Hz fields in the face of suppression of the evidence on the part of power companies, US armed forces, health departments, and the Electric Power Research Institute, the research organ of the electricity industry.
2. Stanley Aronowitz, *Science as Power: Discourse and Idealogy in Modern Society* (Minneapolis: University of Minnesota Press, 1988).
3. The official organ of CSICOP is *The Skeptical Inquirer*, while its main outlet for book publications is Prometheus Books, 700 East Amherst St, Buffalo, NY 14215.

4. Quoted in Arthur Wrobel, ed., *Pseudo-Science and Society in Nineteenth-Century America* (Lexington: University of Kentucky Press, 1987), p. 13. This collection of essays, indispensable to a historical understanding of the demarcation debate, emphasizes the rationalist, reformist, egalitarian, and utilitarian assumptions that lay behind the pseudosciences.
5. Christopher Hill, *The World Turned Upside Down* (New York: Viking, 1972).
6. Morris Berman, *The Reenchantment of the World* (Ithaca: Cornell University Press, 1981), pp. 89–90.
7. See Meaghan Morris, "Banality in Cultural Studies," in Patricia Mellencamp, ed., *The Logics of Television* (Bloomington: Indiana University Press, 1990), pp. 14–43, for a critique of this tendency in cultural studies.
8. Donna Haraway, "Situated Knowledges: The Science Question in Feminism and the Privilege of Partial Perspective," in *Simians, Cyborgs, and Women: The Reinvention of Nature* (New York: Routledge, 1991), p. 196.
9. Michael Hutchison, *Megabrain: New Tools and Techniques for Brain Growth and Mind Expansion* (New York: William Morrow, 1986), p. 17.
10. Ibid., p. 29.
11. Ibid., p. 111.
12. See Leslie Patten with Terry Patten, *Biocircuits: Amazing New Tools for Energy Health* (Tiburon, CA: H.J. Kramer, 1988), a book that makes the case for Eeman screens, manufactured by the Pattens themselves.
13. For a modern medical attempt to reintroduce the case for vitalism, see Robert O. Becker and Gary Selden, *The Body Electric: Electromagnetism and the Foundation of Life* (New York: William Morrow, 1985).
14. Rosalind Coward has suggested that the focus of holistic therapies on "energy" also reflects the general form of industrial ideology's reliance on concepts of productivity and efficiency: "When it is applied to the body, energy appears as a metaphor for the kinds of productive relations which individuals have to an advanced industrial society." *The Whole Truth: The Myth of Alternative Health* (London: Faber & Faber, 1989), p. 57. This tendency is even more evident in many of the New Age interests in fringe science; the obsession with the work of forgotten or neglected scientist–inventors like Nicolas Tesla; or the dedicated amateur research devoted to producing more efficient techniques of energy production, driven on by a neo-alchemical pursuit of "free energy," or perpetual motion machines. See, for example, *The Journal of Borderland Research* published by Borderland Sciences Research Foundation, PO Box 429, Garberville, CA 95440; or the infolios produced by *Rex Research*, PO Box 1258, Berkeley, CA 94701.
15. Patten and Patten, p. 159.
16. Jon Klimo, *Channeling: Investigations on Receiving Information from Paranormal Sources* (Los Angeles: Jeremy Tarcher, 1987), pp. 72–3.
17. Don Elkins, Carla Rueckert, and James Allen McCarty, *The RA Material: An Ancient Astronaut Speaks* (Norfolk, VA: Donning Company, 1984), p. 53.
18. See Douglas Curran's folk-ethnographic survey, *In Advance of the Landing* (New York: Abbeville Press, 1985).

19. See the essays collected in Ken Wilber, ed., *The Holographic Paradigm and Other Paradoxes: Exploring the Leading Edge of Science* (Boston and London: Shambala, 1985).
20. In his influential book, *A New Science of Life* (Los Angeles: J.P. Tarcher, 1981), Rupert Sheldrake puts the case for what he calls M-fields—morphogenetic fields not yet discovered— operating on a subquantum level, linking every pattern in the universe.
21. Ken Wilber, ed., *Quantum Questions: Mystical Writings of the World's Greatest Physicists* (Boston: Shambhala, 1984).
22. Fritjof Capra, *The Tao of Physics: An Exploration of the Parallels Between Modern Physics and Eastern Mysticism* (Boston: Shambhala, 1975).
23. Marilyn Ferguson, *The Aquarian Conspiracy: Personal and Social Transformation in Our Time* (Los Angeles: J.P. Tarcher, 1987; 1st edn. 1980), p. 69.
24. Ramtha and Douglas Mahr, *Destination Freedom: A Time-Travel Adventure. Stage II: Arrival Instruction* (New York: Prentice Hall, 1989), back cover.
25. Robert Pirsig, *Zen and the Art of Motorcycle Maintenance* (New York: William Morrow, 1974), p. 16.
26. Ilya Prigogine and Isabelle Stengers, *Order Out of Chaos: Man's New Dialogue with Nature* (New York: Bantam, 1984).
27. José Argüelles, *The Transformative Vision: Reflections on the Nature and History of Human Expression* (Berkeley: Shambhala, 1975), pp. 277–87.
28. See J.A. English-Hueck's ethnographic study of the holistic practitioners in Paraiso, California, *Health in the New Age: A Study in California Holistic Practices* (Albuquerque: University of New Mexico Press, 1990).
29. Ivan Illich, *Medical Nemesis: The Expropriation of Health* (New York: Penguin, 1976).
30. André Gorz, *Ecology as Politics*, trans. Patsy Vigderman and Jonathan Cloud (Boston: South End Press, 1980), p. 152.
31. Ferguson, *The Aquarian Conspiracy*, p. 257.
32. Rosalind Coward, *The Whole Truth: The Myth of Alternative Health*, pp. 197–8. Coward's incisive analysis is less tolerant of the holistic movement than my own critique. She is generally less willing to extend any political value to the cultural or symbolic *alternatives* offered by New Age practices.

    In addition, her book is primarily an account of the British movement. Any comparative account of the differences between the British and the North American movements would have to address different national factors and influences. In the North American case—the awesome economic power of a professional body like the American Medical Association (next to the military, medicine is the largest recipient of legal profit in the United States); the historical role of "snake-oil" entrepreneurialism in frontier culture; and thirdly, the pesistence of revivalist, millenialist, and utopian elements in North American religious culture, which exert a powerful political pressure upon social movements at particular times. In the British case—the hegemonic presence of a welfare state and all of the social functions that are paternalistically adopted by such a system; and the history of the royal family's influential fondness for homeopathic medicine, reinforced by the aristocracy's dislike for modernity in general. There may be grounds for assuming that, historically, the British cults of Nature are more socially stratified than their North American populist counterpart.

33. Witold Rybczynski notes that "differences in posture, like differences in eating utensils . . . divide the world as profoundly as political boundaries. Regarding posture there are two camps: the sitters-up (the so-called western world) and the squatters (everyone else). Although there is no Iron Curtain separating the two sides, neither feels comfortable in the position of the other . . . . Why have certain cultures adopted a sitting-up posture when others did not? There seems to be no satisfactory answer to this apparently simple question. It is tempting to suggest that furniture was developed as a functional response to cold floors, and it is true that most of the squatting world is in the tropics. But the originators of sitting furniture—the Mesopotamians, the Egyptians, and the Greeks—all lived in warm climates." *Home: A Short History of an Idea* (New York: Viking, 1986), pp. 78–9. Unfortunately, Rybczynski does not extend his comments to the cultural divisions within the world of sanitary arrangements.
34. Laurence Foss and Kenneth Rothenberg, *The Second Medical Revolution: From Biomedicine to Infomedicine* (Boston: Shambala, 1988).
35. Bruno Latour, *Science in Action: How to Follow Scientists and Engineers Through Society* (Cambridge, Mass: Harvard University Press, 1987), p. 174.
36. Sagan believes that science will win out every time in the holy war with pseudoscience because "to whatever measure this term has any meaning, science has the additional virtue, and it is not an inconsiderable one, of being true." "The Burden of Skepticism," *Not Necessarily the New Age*, ed. Robert Basil (Buffalo: Prometheus Books, 1988), p. 372.
37. Sagan, "The Burden of Skepticism," p. 361.
38. Maureen O'Hara, "Of Myths and Monkeys," in Ted Schultz, ed., *The Fringes of Reason* (New York: Harmony Books, 1989), p. 184.
39. See O'Hara above, and O'Hara, "Science, Pseudoscience and Mythmongering," in *Not Necessarily the New Age*, pp. 145–64. O'Hara is a leading figure in the humanistic psychology movement, which emerged as a democratizing challenge to behaviorist orthodoxies in psychology.
40. In his book, *The Mayan Factor: Path Beyond Technology* (Santa Fe: Bear & Co., 1987), José Argüelles proposed the thesis that the completion of the ancient Mayan calendar on August 16/17, 1987, would generate a massive leap in global consciousness, reversing, among other things, the "resonant field paradigm" of modern rationalism which began in 1618 with Descartes's publication of the *Meditations*. On that date, Argüelles, in his rich high-tech vernacular, predicted the triumph of the New Age case for the "path beyond technology":

> As the index of the rate of planetary acceleration, technology will indeed have transformed itself. Through synchronization, this transformation will show us that all of our electro-magnetic hardware and galactic light-body programming, it is we ourselves, Maya returned, who in our own bodies are the best and most sophisticated technology there is—the path beyond technology.

Argüelles's rhetoric—the Harmonic Convergence was to be "the exponential acceleration of the wave harmonic of history as it phases into a moment of unprecedented synchronization"—is a good example of New Age rhetoric's exploitation of high-tech scientific languages.

41. Ferguson, *The Aquarian Conspiracy*, p. 23.
42. The most adequate example of an attempt to offer a systematic overview of New Age politics is Mark Satin, *New Age Politics: Healing, Self and Society* (New York: Delta, 1979).
43. Bill Devall and George Sessions, *Deep Ecology: Living As If Nature Mattered* (Layton, Utah: Gibbs Smith, 1985), pp. 6, 141.
44. Jürgen Habermas, "Modernity—An Incomplete Project?" in Hal Foster, ed., *The Anti-Aesthetic: Essays on Postmodern Culture* (Port Townsend, Washington: Bay Press, 1983), pp. 3–15.

CHAPTER 2

1. John Markoff, *The New York Times*, May 30, 1989.
2. Bryan Kocher, "A Hygiene Lesson," *Communications of the ACM*, 32, 1 (January 1989), p. 3.
3. Jon A. Rochlis and Mark W. Eichen, "With Microscope and Tweezers: The Worm from MIT's Perspective," *Communications of the ACM*, 32, 6 (June 1989), p. 697.
4. Philip Elmer-DeWitt, "Invasion of the Body Snatchers," *Time* (September 26, 1988), pp. 62–7.
5. Judith Williamson, "Every Virus Tells a Story: The Meaning of HIV and AIDS," *Taking Liberties: AIDS and Cultural Politics*, eds. Erica Carter and Simon Watney (London: Serpent's Tail/ICA, 1989), p. 69.
6. "Pulsing the system" is a well-known intelligence process in which, for example, planes deliberately fly over enemy radar installations in order to determine what frequencies they use and how they are arranged. It has been suggested that Morris Sr and Morris Jr worked in collusion as part of an NSA operation to pulse the Internet system, and to generate public support for a legal clampdown on hacking. See Allan Lundell, *Virus! The Secret World of Computer Invaders That Breed and Destroy* (Chicago: Contemporary Books, 1989), pp. 12–18. As is the case with all such conspiracy theories, no actual conspiracy need have existed for the consequences—in this case, the benefits for the intelligence community—to have been more or less the same.
7. For details of these raids, see *2600: The Hacker's Quarterly*, 7, 1 (Spring 1990).
8. "Hackers in Jail," *2600: The Hacker's Quarterly*, 6, 1 (Spring 1989), pp. 22–3. The recent Secret Service action that threatened to shut down *Phrack*, an electronic newsletter operating out of St Louis, confirms *2600*'s thesis; nonelectronic publication would not be censored in the same way.
9. This is not to say that the new laws cannot themselves be used to protect hacker institutions however. *2600* has advised operators of bulletin boards to declare them private property, thereby guaranteeing protection under the Electronic Privacy Act against unauthorized entry by the FBI.
10. Hugo Cornwall, *The Hacker's Handbook*, 3rd edn (London: Century, 1988) 181, pp.

2–6. In Britain, for the most part, hacking is still looked upon as a matter for the civil, rather than the criminal, courts.

11. Discussions about civil liberties and property rights, for example, tend to preoccupy most of the participants in the electronic forum published as "Is Computer Hacking a Crime?" in *Harper's*, 280, 1678 (March 1990), pp. 45–58.
12. See Hugo Cornwall, *Data Theft* (London: Heinemann, 1987).
13. Bill Landreth, *Out of the Inner Circle: The True Story of a Computer Intruder Capable of Cracking the Nation's Most Secure Computer Systems* (Redmond, WA: Tempus, Microsoft, 1989), p. 10.
14. *The Computer Worm: A Report to the Provost of Cornell University on an Investigation Conducted by the Commission of Preliminary Enquiry* (Ithaca, NY: Cornell University, 1989).
15. Ibid., p. 8.
16. A.K. Dewdney, the "computer recreations" columnist at *Scientific American*, was the first to publicize the details of this game of battle programs in an article in the magazine's May 1984 issue. In a follow-up article in March 1985, "A Core War Bestiary of Viruses, Worms, and Other Threats to Computer Memories," Dewdney described the wide range of "software creatures" that readers' responses had brought to light. A third column, in March 1989, was written in an exculpatory mode to refute any connection between his original advertisement of the Core War program and the spate of recent viruses.
17. Andrew Ross, *No Respect: Intellectuals and Popular Culture* (New York: Routledge, 1989), p. 212.
18. The definitive computer liberation book is Ted Nelson's *Computer Lib: Dream Machines* (Redmond, WA: Tempus, 1987, revised edn; original pub. 1974).
19. See Alice Bach's "Phreakers" series, which narrates the mystery-and-suspense adventures of two teenage girl hackers: *The Bully of Library Place* (New York: Dell, 1987), *Double Bucky Shanghai* (New York: Dell, 1987), *Parrot Woman* (New York: Dell, 1987), *Ragwars* (New York: Dell, 1987), and others. The hacker has also appeared recently as a demonized figure in other genres: *The Hacker*, a horror novel by Chet Day (Pocket Books, 1989); and as a new Batman comic series, drawn by Pepe Moreno.
20. John Markoff, "Cyberpunks Seek Thrills in Computerized Mischief," *New York Times* (November 26, 1988), 1, 28.
21. Dennis Hayes, *Behind the Silicon Curtain: The Seductions of Work in a Lonely Era* (Boston: South End, 1989), p. 93.

    One striking historical precedent for the hacking subculture, suggested to me by Carolyn Marvin, was the widespread activity of amateur or "ham" wireless operators in the first two decades of the century. Initially lionized in the press as boy-inventor heroes for their technical ingenuity and daring adventures with the ether, this white middle-class subculture was increasingly demonized by the US Navy (whose signals the amateurs prankishly interfered with), which was crusading for complete military control of the airwaves in the name of national security. The amateurs lobbied with democratic rhetoric for the public's right to access the airwaves, and, although partially successful in their case against the Navy, lost out ultimately to big commercial interests when Congress approved the creation of a broadcasting monopoly after World War I in the form of RCA. See Susan J. Douglas, *Inventing American Broadcasting 1899–1922* (Baltimore: Johns Hopkins University Press, 1987), pp. 187–291.

22. "Sabotage," *Processed World*, 11 (Summer 1984), pp. 37–8.
23. Hayes, *Behind the Silicon Curtain*, p. 98.
24. *Bad Attitudes: The Processed World Anthology*, ed. Chris Carlsson with Mark Leger (London: Verso, 1990), contains highlights from the magazine's first eight years.
25. *The Amateur Computerist*, available from R. Hauben, PO Box 4344, Dearborn, MI 48126.
26. Kevin Robins and Frank Webster, "Athens without Slaves . . . or Slaves without Athens? The Neurosis of Technology," *Science as Culture*, 3 (1988), pp. 7–53.
27. See, for example, the collection of essays edited by Vincent Mosco and Janet Wasko, *The Political Economy of Information* (Madison: University of Wisconsin Press, 1988); Kevin Robins and Frank Webster, *Information Technology: A Luddite Analysis* (Norwood, NJ: Ablex, 1986).
28. Tom Athanasiou and Staff, "Encryption and the Dossier Society," *Processed World*, 16 (1986), pp. 12–17.
29. See Kevin Wilson, *Technologies of Control: The New Interactive Media for the Home* (Madison: University of Wisconsin Press, 1988), pp. 121–5.
30. Hayes, *Behind the Silicon Curtain*, pp. 63–80.
31. "Our Friend the VDT," *Processed World*, 22 (Summer 1988), pp. 24–5.
32. See Kevin Robins and Frank Webster, "Cybernetic Capitalism," in *The Political Economy of Information*, ed. Vincent Mosco and Janet Wasko (Madison: University of Wisconsin Press, 1988), pp. 44–75.
33. Hans Magnus Enzensberger, "Constituents of a Theory of the Media," *The Consciousness Industry*, trans. Stuart Hood (New York: Seabury, 1974).
34. Barbara Garson, *The Electronic Sweatshop: How Computers are Transforming the Office of the Future into the Factory of the Past* (New York: Simon & Schuster, 1988), pp. 244–5.
35. See Marike Finlay's Foucauldian analysis, *Powermatics: A Discursive Critique of New Technology* (London: Routledge & Kegan Paul, 1987). A more conventional culturalist argument can be found in Stephen Hill, *The Tragedy of Technology: Human Liberation versus Domination in the late Twentieth Century* (London: Pluto, 1988).

CHAPTER 3

1. William Gibson, "The Gernsback Continuum," in *Mirrorshades: The Cyberpunk Anthology*, ed. Bruce Sterling (New York: Arbor House, 1986), pp. 1–3.
2. Most histories of the genre accept this model. For an interesting example, see Peter Fitting's periodicization of SF history in "The Modern Anglo-American SF Novel: Utopian Longing and Capitalist Cooptation," *Science-Fiction Studies* 6, 1 (March 1979), pp. 59–76. Fitting's brief history of "modern SF" begins, symptomatically, in 1937, *after* Gernsbackianism, and consistently "looks forward" to the heyday of feminist utopian writing in the seventies. In a response to this article, Samuel Delany

discusses the earlier Gernsbackian period, but only to characterize it as "uncritically accepting technology," or else to praise what later Campbellian writers took from "the nascent critique of the philosophy of science that they had found in a fraction of the SF from the 1920s and '30s." "Reflections on Historical Models in Modern English Language Science Fiction," *Science Fiction Studies*, 7, 2 (July 1980), pp. 135–49, 140.

3. Brian W. Aldiss (assisted by David Wingrove), *Trillion Year Spree: The History of Science Fiction* (New York: Atheneum, 1986), pp. 175–7.
4. Antonio Gramsci, "Americanism and Fordism," *Selections from the Prison Notebooks*, trans. and ed. Quintin Hoare and Geoffrey Nowell-Smith (New York: International Publishers, 1971), pp. 279–318.
5. Tony Goodstone, ed., *The Pulps: Fifty Years of American Pop Culture* (New York: Chelsea House, 1970), Preface.
6. Isaac Asimov, ed., *Before the Golden Age* (Garden City: Doubleday, 1974), p. 15.
7. See Michael Ashley, ed., *The History of the Science Fiction Magazine*, Part 1, *1926–35* (London: New English Library, 1974), pp. 11–49.
8. Sam Moskowitz, *Explorers of the Infinite: Shapers of Science Fiction* (Cleveland and New York: World Publishing Company, 1960), p. 226.
9. Susan Douglas tells this story in *Inventing American Broadcasting, 1899–1920* (Baltimore: Johns Hopkins University Press, 1987), pp. 144ff.
10. Patrick Parrinder, ed., *H.G. Wells: The Critical Heritage* (London: RKP, 1972), pp. 101–2.
11. See Paul A. Carter, *The Creation of Tomorrow: Fifty Years of Magazine Science Fiction* (New York: Columbia University Press, 1977), especially pp. 1–28.
12. Sam Moskowitz, *Explorers*, pp. 313–33.
13. *Amazing Stories*, 1, 2 (May 1926), p. 99.
14. Manfred Nagel, "SF, Occult Sciences and Nazi Myths," *Science-Fiction Studies*, 1, 3 (Spring 1974), pp. 185–97; and William B. Fischer, *The Empire Strikes Out: Kurd Lasswitz, Hans Dominik, and the Development of German Science Fiction* (Bowling Green: Bowling Green University Popular Press, 1984).
15. Alan Huntingdon has dealt with these questions at some length in *Rationalizing Genius: Ideological Strategies in the Classic American Science Fiction Short Story* (New Brunswick: Rutgers University Press, 1989).
16. For an account of the relation of the SF pulp style to Streamline Moderne, see Kathleen Church Plummer, "The Streamlined Moderne" in *Art in America*, vol. 62, no. 1 (Jan–Feb 1974), pp. 46–54. See also Donald J. Bush, *The Streamlined Decade* (New York: George Braziller, 1975); and John Perreault, *Streamline Design: How the Future Was* (Flushing, NY: Queens Museum, 1984).
17. Brian Ash, *Faces of the Future: The Lessons of Science Fiction* (New York: Taplinger, 1975), pp. 67–8.
18. Sam Moskowitz, *The Immortal Storm: A History of Science Fiction Fandom* (Atlanta: Atlanta Science Fiction Organization Press, 1954), p. 160. While Moskowitz's book is the most exhaustive documentary of the period of fandom, Moskowitz himself was a leading figure in New Fandom, the fan organization that was politically opposed to the Futurians, and so his book is a rather personal record of the history.

19. Tremaine, for example, warned Asimov, one of the culprits, that anyone who followed this practice "ought to be blacklisted." Damon Knight, *The Futurians* (New York: John Day, 1977), p. 90. While Knight, a late Futurian member, plays down the political activities of the group, he provides the best account of their communal living style, cohabiting in a succession of apartments where Communist literature shared the shelves with pornography. Close-knit and hyperactive, they were raided by the FBI for having a printing press, and were raided again later by the vice squad on suspicion of being gay. Frederik Pohl's memoir, *The Way the Future Was: A Memoir* (New York: Ballantine, 1978), is by and large a more faithful account of the politics of the Futurians.
20. Knight, p. 66.
21. William Sims Bainbridge, *Dimensions of Science Fiction* (Cambridge, MA: Harvard University Press, 1986), p. 207.
22. See Cecelia Tichi's discussion of the romantic cult of the engineer in *Shifting Gears: Technology, Literature, Culture in Modernist America* (Chapel Hill: University of North Carolina Press, 1987).
23. For two histories of the movement, see Henry Elsner Jr, *The Technocrats: Prophets of Automation* (Syracuse: Syracuse University Press, 1967); and William E. Akin, *Technocracy and the American Dream: The Technocrat Movement, 1900–1941* (Berkeley: University of California Press, 1977).
24. Howard Scott, "Technocracy Speaks," in Thomas Parke Hughes, ed., *Changing Attitudes Towards American Technology* (New York: Harper & Row, 1975), p. 305.
25. Ibid., p. 300.
26. Technocracy preached a "rendez-vous with destiny," whose continuity with the imperialist tradition of Manifest Destiny was underscored by the fact that plans for the Technate included Canada, and, at times, extended to Mexico. See Howard Scott, "A Rendez-Vous with Destiny," In Carroll Pursell Jr, ed., *Readings in Technology and American Life* (New York: Oxford University Press, 1969), pp. 351–7.

    As for the anti-humanist elements in Scott's program, the movement itself was soon divided between Scott's Technocracy Inc. and the more humanistically oriented Continental Congress on Technocracy. Harold Loeb, associated with the latter, wrote the most fully social account of a humanist utopia in *Life in a Technocracy* (New York: Viking, 1933).
27. Edwin T. Layton Jr, *The Revolt of the Engineers: Social Responsibility and the American Engineering Profession* (Cleveland: Press of Case Western Reserve University, 1971), p. 277.
28. Tichi, pp. 19–25.
29. See Thorstein Veblen's "A Memorandum on a Practicable Soviet of Technicians," in *The Engineers and the Price System* (New York: Viking, 1921).
30. David Noble, *America By Design: Science, Technology and the Rise of Corporate Capitalism* (New York: Knopf, 1977), pp. 4–5.
31. 31% of high school boys in 1922 said that they wanted to be engineers; agriculture came in second with 24%. Tichi, p. 103.
32. Jack Williamson, "The Legion of Time," *Astounding Stories* (June 1938), p. 38.

33. Bruce Franklin argues this point thoroughly in his book, *Robert A. Heinlein: America as Science Fiction* (New York: OUP, 1980)
34. Joseph Custer, "The World of Tomorrow: Science, Culture and Community at the New York World's Fair," in *Dawn of a New Day: The New York World's Fair, 1939/40*, curated by Helen A. Harrison (New York: Queens Museum/New York University Press, 1980), pp. 4–5.
35. Bruce Franklin has compared the uncritical enthusiasm of the "verbal and model worlds projected by the large corporations" at the 1939 World's Fair with the more dystopian sentiments found in stories published in *Astounding Stories* in the same year. "America as Science Fiction: 1939," George Slusser, Eric Rabkin, and Robert Scholes, eds, *Coordinates: Placing Science Fiction and Fantasy* (Carbondale: Southern Illinois University Press, 1983), pp. 107–23.
36. Albert Einstein, "The Five Thousand-Year Time Capsule of the Westinghouse Electric Company," *Science*, 88 (September 1938), p. 275.
37. In his admirable history of the politicization of scientists in the thirties, Peter J. Kuznick's *Beyond the Laboratory: Scientists as Political Activists in 1930s America* (Chicago: University of Chicago Press, 1987) dispels the notion that the scientific community only became socially organized after the war around the issue of the hydrogen bomb. Kuznick discusses scientists' activist roles in the mid to late thirties in the science and society movements, the medical reform movement, the civil liberties movement, the anti-fascist movement, and the Spanish civil war. He also recounts the history of progressive professional organizations like the American Association for the Advancement of Science and the American Association of Scientific Workers and their support for attempts to direct scientific research and development towards public ends.
38. Franklin, p. 121.
39. Lewis Mumford, *Technics and Civilization* (New York: Harcourt, Brace and Co., 1934), p. 248.
40. Ibid., p. 250.
41. See James Carey (with John J. Quirk), "The Mythos of the Electronic Revolution," in *Communication and Culture: Essays on Media and Society* (Boston: Unwin & Hyman, 1989), pp. 113–41.

CHAPTER 4

1. The polluting capacity of the postwar automobile was different in kind from prewar models that had spewed carbon monoxide alone. Larger cylinders and higher compression ratios in the new overpowered American engines meant that the greater engine heat combined nitrogen and oxygen to form nitrous oxide. The result: ozone, petrochemical smog, and acidic nitrates in the air.
2. Sheldon J. Reaven, "New Frontiers: Science and Technology at the Fair," in *Remembering the Future: The New York World's Fair from 1939 to 1964* (NY: Queens Museum/Rizzoli, 1989), p. 98.

3. Morris Dickstein, "From the Thirties to the Sixties: The New York World's Fair in its Own Time," in *Remembering the Future*, pp. 32–4.
4. Footage, on the ground, shot by Akiri Iwasaki in Hiroshima and Nagasaki was impounded by the Pentagon, and appeared a quarter of a century later in *Hiroshima-–Nagasaki: August 1945*, a documentary put together at Columbia University. See Erik Barnouw's account," The Hiroshima–Nagasaki Footage: A Report, "*Historical Journal of Film Radio and Television*, II, 1 (March 1982).
5. Bruce Sterling, "Just a Sci-Fi Guy," *Locus: The Newspaper of the Science Fiction Field*, 328 (May 1988), p. 6.
6. For an account of some of the official techno-utopian fantasies woven around the benefits of the atom, see Paul Boyer, *By the Bomb's Early Light: American Thought and Culture at the Dawn of the Atomic Age* (New York: Pantheon, 1985).
7. Edward Teller (with Allen Brown), *The Legacy of Hiroshima* (Garden City: Doubleday, 1962), p. 81.
8. Office of Technology Assessment, *The Effects of Nuclear War* (Washington: OTA, 1979).
9. SF survivalist literature is a tradition of adventure fiction with specifically New World overtones. By contrast, British fiction in the same mold, as David Dowling has argued, tended to focus on the theological or moral significance of the apocalyptic events themselves. David Dowling, *Fictions of Nuclear Disaster* (Iowa City: University of Iowa Press, 1987), p. 62.
10. See Tom Moylan, *Demand the Impossible: Science Fiction and the Utopian Imagination* (New York: Methuen, 1986).
11. Fredric Jameson, "Progress or Utopia, or, Can We Imagine the Future?" *Science-Fiction Studies*, 9, 2 (July 1982), pp. 147–58; "Nostalgia for the Present," *South Atlantic Quarterly*, 88, 2 (1989), pp. 517–37.
12. See Fred Glass, "The 'New Bad Future': *Robocop* and 1980s Sci-Fi Films," *Science as Culture*, 5 (1989), pp. 7–49.
13. Istvan Csicsery-Ronay, "Cyberpunk and Neuromanticism," *Mississippi Review*, 16, 2/3 (1988), 266-78, p. 267.
14. Rudy Rucker, *Mississippi Review*, 16, 2/3 (1988), p. 57.
15. See my "Cowboys, Cadillacs and Cosmonauts: Families, Film Genres, and Technocultures," in Joseph Boone and Michael Cadden, eds, *Engendering Men: The Question of Male Feminist Criticism* (New York: Routledge, 1990) pp. 87–101.
16. Page numbers of William Gibson's books refer to the following: *Burning Chrome* (New York: Arbor House, 1986); *Neuromancer* (New York: Ace, 1984); *Count Zero* (New York: Arbor House, 1986); *Mona Lisa Overdrive* (New York: Bantam, 1988).
17. Peter Fitting, "The Lessons of Cyberpunk," in Constance Penley and Andrew Ross, eds, *Technoculture* (Minneapolis: University of Minnesota Press, 1991); Tom Moylan, "Run with the Home Boys: Cyberpunk SF and Postmodern Agency" (unpublished ms.); and Csicsery-Ronay, "Cyberpunk and Neuromanticism."
18. Pam Rosenthal, "Jacked In: Fordism, Cyberpunk, Marxism," *Socialist Review* (Spring 1991) pp. 79-104, p. 100.
19. Takayuki Tatsumi, "Eye to Eye: An Interview with William Gibson," *Science Fiction EYE*, I, 1 (1987), 6–17, p. 12.

20. The lesbian detective genre is a notable antidote to what Mike Davis calls "the emergence of *homo reaganus* in postmodern noir." Of James Ellroy's novels, he writes: "The result feels very much like the actual moral texture of the Reagan–Bush era: a supersaturation of corruption that fails any longer to outrage or even interest." Mike Davis, *City of Quartz: Excavating the Future in Los Angeles* (London: Verso, 1990), p. 45.
21. Bruce Sterling, *Islands in the Net* (New York: William Morrow, 1988), p. 234.
22. The *Daredevil* adventures of Elektra ran from 1980 to 1984, and are collected in the graphic novel *The Elektra Saga* (Marvel, 1989). The original run of the *Elektra Assassin* series was published by Marvel (August 1986–March 1987).
23. Acker's throwaway sketch of this future scenario is flip but telling: "Except for Manhattan, which had been left to the rich, all of the eastern American urban centers had been left to the packs of wild dogs, wild cats and blacks who lived in and under the streets. There were no more whites there except for gays." *Empire of the Senseless* (New York: Grove Press, 1988), p. 36.
24. "View From the Edge: The Cyberpunk Handbook," in *Cyberpunk: The Role-Playing Game of the Dark Future* (R. Talsorian Games, Inc., 1988), p. 2.
25. Donna Haraway, "A Manifesto for Cyborgs: Science, Technology and Socialist Feminism in the 1980s," *Socialist Review*, 15, 2 (1985), 65–107.
26. Queen Mu and R.U. Sirius, editorial, *Mondo 2000*, 7 (Fall 1989). Leary has taken the cyberpunks to his bosom. See his conversation with William Gibson, "High Tech High Life," in the same issue, pp. 58–66.
27. R.U. Sirius, "The New Species Comes of Age," *High Frontiers*, 4 (1987), p. 6.
28. Two examples of this genre are Stewart Brand, *The Media Lab: Inventing the Future at M.I.T.* (New York: Viking Penguin, 1987); and Grant Fjermedal, *The Tomorrow Makers: A Brave New World of Living-Brain Machines* (New York: Macmillan, 1986). Brand eulogizes the creativity of the Media Lab while noting that "MIT is more merrily in bed with industry and government than any other academic institution in the world" (p. 5). Fjermedal does a breathless tour of the whiz-kids of artificial intelligence, symptomatically visiting the Harvard Divinity School at the same time as MIT to underline the book's thesis that these scientists are all "trying to play God."
29. K. Eric Drexler, *Engines of Creation: The Coming Age of Nanotechnology* (New York: Doubleday, 1986), p. 41.
30. Rudy Rucker's concept of cellular automata, developed at Autodesk, is an experiment in artificial life systems, as opposed to artificial intelligence. Rather than dictate what intelligence ought to be, from above, like the AI people, he favors allowing a behaviorally rich computer environment to evolve on its own. Much of the anarchist spirit behind this idea can be found in his novels *Wetware* (1982) and *Software* (1988), which describe the revolt of robot labor on the Moon, and the creation of an anarchist robot society. The robots, or boppers, routinely download their intelligence into new bodies.
31. Fjermedal, p. 33.
32. Weizenbaum's own remarks, in a conversation with Fjermedal, in Fjermedal, p. 136.
33. Murray Bookchin, *Remaking Society: Pathways Toward a Green Future* (Boston: South End Press, 1990).

CHAPTER 5

1. Daniel Bell, "The Year 2000: The Trajectory of an Idea," in Bell, ed., *Toward the Year 2000: Work in Progress* (Boston: Houghton Mifflin, 1968), p. 3.
2. Ibid., p. 8.
3. The World Future Society, *An Introduction to the Study of the Future: Resources Directory for America's Third Century, Part One* (Washington: World Future Society, 1977), pp. 130–2.
4. Arthur Waskow, *Running Riot: A Journey Through the Official Disasters and Creative Disorder in American Society* (New York: Herder & Herder, 1970), p. 138.
5. Lewis Mumford, *Technics and Civilization*, p. 23.
6. Quoted in Cedric B. Cowing, *Populists, Lungers and Progressives: A Social History of Stock and Commodity Speculation, 1890–1936* (Princeton: Princeton University Press, 1965), p. 253.
7. Georgi Shakhnazarov, *Futurology Fiasco: A Critical Study of Non-Marxist Concepts of How Society Develops*, trans. Vic Scheierson (Moscow: Progress Publishers, 1982), p. 7.
8. Fred Polak, *The Image of the Future: Enlightening the Past, Orientating the Present, Forecasting the Future*, trans. Elise Boulding (Leyden: A.W. Sijthoff/New York: Oceana, 1961), 2 vols.
9. Donella Meadows et al., *The Limits to Growth: A Report for the Club of Rome's Project on the Predicament of Mankind* (New York: Universe Books, 1972), p. 196.
10. Carl Sagan and Richard Turco, *A Path Where No Man Thought: Nuclear Winter and the End of the Arms Race* (New York: Random House, 1991).
11. Council of Environmental Quality and US State Department, *The Global 2000 Report to the President of the U.S.: Entering the 21st Century Volume 1: the Summary Report* (New York: Pergamon, 1980), Gerald Barney, Study Director, in the Preface.
12. *Averting Catastrophe: The Global Challenge.* Report of the Subcommittee on International Economics of the Joint Economic Committee, Congress of the United States, October 10, 1980 (Washington: US Government Printing Office, 1980), p. iii.
13. David Dickson exhaustively describes this history in *The New Politics of Science* (Chicago: University of Chicago Press, 1988).
14. World Commission on Environment and Development, *Our Common Future* (New York: Oxford University Press, 1987); and Linda Starke, *Signs of Hope: Working Towards Our Common Future* (New York: Oxford University Press, 1990).
15. Murray Bookchin, *Remaking Society: Pathways to a Green Future* (Boston: South End Press, 1990), p. 93.
16. Raymond Williams, *The Year 2000: A Radical Look at the Future—and What We Can Do to Change It* (New York: Pantheon, 1983), pp. 244–7.
17. *New Left Review*, 1 (1960), p. 1.
18. For a fuller analysis of the consequences of this kind of politics, see the essays collected in Stuart Hall and Martin Jacques, eds, *New Times: The Changing Face of Politics in the 1990s* (London: Lawrence & Wishart, 1989).

CHAPTER 6

1. Who can forget Ray Bradbury's short story, "The Sound of Thunder," an SF biological variant of this witticism, in which a time-traveller accidentally kills a butterfly in the remote past, and returns to find the present quite altered as a result?
2. Thomas Jefferson, "Jefferson's Summary of his Meteorological Journal for the Years 1810 through 1816 at *Monticello*," in *Thomas Jefferson's Garden Book*, annotated by Edwin Morris Betts (Philadelphia: American Philosophical Society, 1944), pp. 622–4.
3. A.A. Miller, "The Use and Misuse of Climatic Resources," *Advancement of Science*, 13, p. 58.
4. Michael Oppenheimer and Robert H. Boyle, *Dead Heat: The Race Against the Greenhouse Effect* (New York: Basic Books, 1990), p. 18.
5. Philip K. Dick, *Do Androids Dream of Electric Sheep?* (New York: Ballantine, 1968), p. 54.
6. See Rhys Carpenter, *Discontinuity in Greek Civilization* (New York: Norton, 1968).
7. Emmanuel Le Roy Ladurie has examined the history of glacial shifts in medieval Europe in *Times of Feast, Times of Famine: A History of Climate Since the Year 1000*, trans. Barbara Bray (Garden City, New Jersey: Doubleday, 1971).
8. The pioneering work on climate change was compleed by H.H. Lamb at the University of East Anglia's Climatic Research Unit, collected in *Climate: Past, Present and Future* (New York: Barnes and Noble, 1972), 2 vols. Other, competing schools of philosophy in climatology stemmed from the work of the Soviet scientist M.I. Budyko, and the Princeton climatologist Joseph Smagorinsky. The chief source for theories of global cooling was the work of Reid Bryson and colleagues at the University of Wisconsin in the early seventies. See Reid A. Bryson and Thomas J. Murray, *Climates of Hunger: Mankind and the World's Changing Weather* (Madison: University of Wisconsin Press, 1977).
9. Nigel Calder, *The Weather Machine* (New York: Viking, 1974); Lowell Ponte, *The Cooling* (Englewood Cliffs: Prentice Hall, 1976); John Gribbin, *Forecasts, Famines and Freezes* (New York: Walker, 1976); D.S. Halacy, *Ice or Fire? Can We Survive Climate Change?* (New York: Harper & Row, 1978). Halacy's book really belongs to the cooling earth school, but it also reflects the growing interest in warming theories.
10. The two reports, "A Study of Climatological Research as it Pertains to Intelligence Problems" and "Potential Implications of Trends in the World's Population, Food Production, and Climate" (August 1974), are reprinted in Impact Team, *The Weather Conspiracy: The Coming of the New Ice Age* (New York: Ballantine, 1977).
11. See Lowell Ponte, *The Cooling*, pp. 169–74.
12. UN General Assembly, December 1974, Resolution 3264, "Prohibition of action to influence the environment and climate for military and other purposes incompatible with the maintenance of international security, human well-being and health."
13. By comparison, the British government's comparative alacrity in supporting broad measures to combat the greenhouse effect can perhaps be attributed more to its nationalist phobia about unstoppable "waves" of immigration than to Margaret Thatcher's efforts to coopt green politics. Crispin Tickell, Britain's UN ambassador

at the time and author of *Climatic Change and World Affairs* (1977), was most persuasive in reminding Prime Minister Thatcher of the potential proportions of the "greenhouse refugee" problem: "even if some people and governments wished to seal themselves off from the rest of the world, they could not do so," he pointed out. "In no country or city can the rich fortify themselves for long against the poor." Quoted in Boyle and Oppenheimer, p. 200, from a newspaper article in *The Independent*, entitled "How Warming Could Create Refugee Crisis" (June 6, 1989).

14. See, for example, Larry Ephron, *The End: The Imminent Ice Age and How We Can Stop It!* (Berkeley: Celestial Arts, 1988).
15. John Hamaker's papers are collected in *The Survival of Civilization* with annotations by Donald Weaver (Burlingame, CA: Hamaker-Weaver Publishers, 1982). Hamaker attributes the death of forests to calcium depletion of soil, caused by the long process of demineralization and acidification in the wake of the glacial retreat. He advocates the project of remineralizing the earth's soil with ground-up rock before the glaciers return to do the job for us.
16. W.J. Maunder, a New Zealand meteorological scientist, has argued that the atmosphere, a "variable and elite resource," must be "accepted as an integral part of the management package" for industry and for all important decision-making at corporate and political levels. *The Uncertainty Business; Risks and Opportunities in Weather and Climate* (London: Methuen, 1986); and *The Value of the Weather* (London: Methuen, 1970). The aim of such arguments is quite explicitly to push weather as a "product." As a result of the global warming debate, however, the rhetoric of atmospheric forecasting has seen a boom that far exceeds Maunder's expectations.
17. James Lovelock, *The Ages of Gaia* (New York: Norton, 1988), p. 212.
18. See Murray Bookchin's eloquent attack on de-socializing tendencies in the ecology movement, in *Remaking Society: Pathways to a Green Future* (Boston: South End Press, 1990).
19. Norman Myers and Gaia Ltd. Staff, *Gaia: An Atlas of Planet Management* (New York: Anchor, 1984).
20. Roger Revelle and Hans E. Seuss, "Carbon Dioxide Exchange Between Atmosphere and Ocean and the Question of an Increase of Atmospheric CO_2 During the Past Decades," *Tellus*, 9 (1957), pp. 18–27.
21. W.J. Maunder notes the disparity between Northern cooling in the 1950s and 1960s and warming in the southern hemisphere throughout the same period. *The Uncertainty Business*, p. 78. Stephen Schneider accounts for US cooling in the same period as a local effect, with only minor significance for the overall global temperatures. *Global Warming: Are We Entering the Greenhouse Century?* (San Francisco: Sierra Club Books, 1989), p. 200. More recently, he has suggested that the transient cooling may have been caused by a sharp increase in sulphur emissions from coal- and oil-burning factories and power plants. These emissions generate sulphate particles that act as condensation nuclei for the formation of cloud droplets, brightening clouds and thus increasing their cooling effects. Stephen H. Schneider, "The Changing Climate: A Risky Planetwide Experiment," in Terrell J. Minger, ed., *Greenhouse Glasnost: The Crisis of Global Warming* (New York: Ecco Press/Institute for Resources Management, 1990), p. 126. Schneider's *Global Warming*, and his earlier *The Genesis Strategy:*

Climate and Global Survival (with L.E. Mesirow) (New York: Plenum, 1976), are good guides to the climate debates written by an active participant at the highest levels.

Other information sources about the greenhouse effect that I have found useful include Francesca Lyman et al., *The Greenhouse Trap* (Boston: Beacon Press, 1990); Jim Falk and Kevin Brownlow, *The Greenhouse Challenge* (New York: Viking Penguin, 1989).

22. Mary Douglas, *Implicit Meanings: Essays in Anthropology* (London: Routledge and Kegan Paul, 1975), pp. 234–6.
23. Jody Berland, "Weathering the North: Climate, Colonialism and the Mediated Body," *Provincial Essays*, vol. 8 (1989), p. 37.
24. Zora Neale Hurston, *Their Eyes Were Watching God* (New York: Harper & Row, 1990), pp. 147, 148.
25. Claude Lévi-Strauss, *The Origin of Table Manners*, trans. John and Doreen Weightman (New York: Harper & Row, 1978), p. 507. Lévi-Strauss concludes that the origin of table manners may lie "in deference towards the world—good manners existing precisely in respecting its obligations," rather than in fearful ways of warding off pollution from the external world: "Whereas we think of good manners as a way of protecting the internal purity of the subject against the external impurity of beings and things, in savage societies, they are a means of protecting the purity of beings and things against the impurity of the subject." p. 504.
26. James Pollard Espy, *The Philosophy of Storms* (Boston: Charles Little and James Brown, 1841), p. 495.
27. For a history of rainmaking, see Clark C. Spence, *The Rainmakers: American Pluviculture to World War II* (Lincoln: University of Nebraska Press, 1980).
28. Quoted in Gary Lockhart, *The Weather Companion* (New York: John Wiley, 1988), p. 100.
29. W.E. Knowles Middleton, *A History of the Theories of Rain: And Other Forms of Precipitation* (New York: Franklin Watts, expression of the American character," p. 160. A century later, more "scientific" rainmakers still urged their claims with appeals to a calculus of distributed costs. The last of the great freelance rainmakers, Irving Langmuir, operating an iodide generator in New Mexico in the late 1940s, estimated the cost of his efforts to local citizens: "Assuming the atmosphere to be 5 miles thick, one thus finds that to get a 30 per cent chance of rain within a given area in New Mexico the cost of the silver iodide is only $1 for 4,000 square miles." To complete the estimate on a national scale, to double the United States rainfall, assuming conditions were the same as in New Mexico, Langmuir felt that it would cost only about $200 a week. Quoted in D.S. Halany, *Ice or Fire? Can We Survive Climate Change?* (New York: Harper and Row, 1978), p. 64.
30. Charles Peirce, *Peirce's Statistics of the Weather: A Meteorological Account of the Weather in Philadelphia from January 1 1790 to January 1 1847* (Philadelphia: Lindsay & Blakiston, 1847).
31. Daniel Boorstin, "The Rise of the Average Man," in *The Decline of Radicalism: Reflections on America Today* (New York: Random House, 1969).
32. See Donald R. Whitnah, *A History of the United States Weather Bureau* (Urbana: University of Illinois Press, 1961).

33. Consider, by way of contrast, the instructions given to voluntary observers of the US Signal Service: no interpretive skills were required of the observer, only a willingness to faithfully measure the temperature, humidity, atmospheric pressure, amount and frequency of rain, direction and velocity of wind, and the electrical condition of the atmosphere. *Instructions for Voluntary Observers of the Signal Service, U.S. Army* (Washington: Government Printing Office, 1882).
34. Robert Marc Friedman, *Appropriating the Weather: Wilhelm Bjerknes and the Construction of a Modern Meteorology* (Ithaca: Cornell University Press, 1989), p. 164.
35. Raymond Williams, "Ideas of Nature," in *Problems in Materialism and Culture* (London: Verso, 1980), pp. 67–85.
36. John Edward Weems, *"If You Don't Like the Weather . . . ": Stories of Texas Weather* (Austin: Texas Monthly Press, 1986), p. 61. The drought had forced many of the white settlers to evacuate and return to the East, but for those who survived, it became a moment of pioneer heroism, monumentalized in the state's history of the conquest of nature and territory.
37. Mark Twain, "Speech on the Weather," *The Family Mark Twain* (New York: Harper, 1935), pp. 1109–10.
38. Twain's more famous observation about the weather (attributed to the editor of the *Hartford Courant*) is that "everybody talks about the weather, but no-one does anything about it." This remark took on a new significance in the early stages of the debate about climate change, when Stephen Schneider made the much publicized comment at an AAAS meeting that "nowadays, everybody is doing something about the weather, but nobody is talking about it." Schneider, *Global Warming*, p. 200.
39. Nonetheless, the history of the weather contains many Althusserian moments when "science" interrupts "ideology." The seventeenth-century discovery of barometric variations, for example, ran counter to all doctrines of "common sense." Atmospheric pressure is something for which we have no sensible perceptions, the lived experience of which is the basis of any notion, Marxist or otherwise, of ideology/common sense.
40. Julius Fast, *Weather Language* (New York: Wyden Books, 1979), pp. 7–8.
41. Benedict Anderson, *Imagined Communities: Reflections on the Origin and Spread of Nationalism* (London: Verso, 1985).
42. Quoted from the Weather Channel's *Weather Guide Calendar*, 1990 (Denver: Accord Publishing, 1989).
43. *American Weather Observer*, 7, 6 (June 1990), p. 2.
44. A more serious ethnographic study of the way in which people are conceptualizing the greenhouse effect can be found in Willett Kempton, "Lay Perspectives on Global Climate Change," Report No. 251, Center for Energy and Environmental Studies, Princeton University, August 1990. Kempton's study shows that most informants had heard of the greenhouse effect but had fundamental misconceptions, from a scientific perspective, of its causes, especially its relation to personal energy consumption.

INDEX

THE HAYMARKET SERIES

Already Published

MARXISM IN THE UNITED STATES: Remapping the History of the American Left *by Paul Buhle*

FIRE IN THE AMERICAS: Forging a Revolutionary Agenda *by Roger Burbach and Orlando Núñez*

THE FINAL FRONTIER: The Rise and Fall of the American Rocket State *by Dale Carter*

CORRUPTIONS OF EMPIRE: Life Studies and the Reagan Era *by Alexander Cockburn*

THE SOCIAL ORIGINS OF PRIVATE LIFE: A History of American Families, 1600–1900 *by Stephanie Coontz*

CITY OF QUARTZ: Excavating the Future in Los Angeles *by Mike Davis*

PRISONERS OF THE AMERICAN DREAM: Politics and Economy in the History of the US Working Class *by Mike Davis*

MECHANIC ACCENTS: Dime Novels and Working-Class Culture in America *by Michael Denning*

POSTMODERNISM AND ITS DISCONTENTS: Theories, Practices *Edited by E. Ann Kaplan*

RANK-AND-FILE REBELLION: Teamsters for a Democratic Union *by Dan La Botz*

AN INJURY TO ALL: The Decline of American Unionism *by Kim Moody*

YOUTH, IDENTITY, POWER: The Chicano Movement *by Carlos Muñoz, Jr.*

ANOTHER TALE TO TELL: Politics and Narrative in Postmodern Culture *by Fred Pfeil*

THE WAGES OF WHITENESS: Race and the Making of the American Working Class *by David R. Roediger*

OUR OWN TIME: A History of American Labor and the Working Day *by David R. Roediger and Philip S. Foner*

THE RISE AND FALL OF THE WHITE REPUBLIC: Class Politics and Mass Culture in Nineteenth-Century America *by Alexander Saxton*

POSTMODERN GEOGRAPHIES: The Reassertion of Space in Critical Social Theory *by Edward W. Soja*

BLACK MACHO AND THE MYTH OF THE SUPERWOMAN *by Michele Wallace*

INVISIBILITY BLUES: From Pop to Theory *by Michele Wallace*

CULTURE WARS: Intellectuals, Politics and Pop by *Jon Wiener*

THE LEFT AND THE DEMOCRATS *The Year Left 1*

TOWARDS A RAINBOW SOCIALISM *The Year Left 2*

RESHAPING THE US LEFT: Popular Struggles in the 1980s *The Year Left 3*

FIRE IN THE HEARTH: The Radical Politics of Place in America *The Year Left 4*

Forthcoming

GENDER AND CLASS *by Johanna Brenner*

ENCOUNTERS WITH THE SPHINX: Journeys of a Radical in Changing Times *by Alexander Cockburn*

THE MERCURY THEATER: Orson Welles and the Popular Front *by Michael Denning*

THE POLITICS OF SOLIDARITY: Central America and the US Left *by Van Gosse*

JAMAICA 1945–1984 *by Winston James*

BLACK AMERICAN POLITICS: From the Washington Marches to Jesse Jackson (Second Edition) *by Manning Marable*

THE OTHER SIDE: Los Angeles from Both Sides of the Border *by Ruben Martinez*

THE SOCIALIST REVIEW READER